CSR, Sustainability, Ethics & Governance

Series Editors

Samuel O. Idowu, London, United Kingdom

René Schmidpeter, Cologne, Germany

For further volumes:
http://www.springer.com/series/11565

Mara Manente • Valeria Minghetti •
Erica Mingotto

Responsible Tourism and CSR

Assessment Systems for Sustainable Development of SMEs in Tourism

Springer

Mara Manente
Valeria Minghetti
Erica Mingotto
CISET
Venice
Italy

ISSN 2196-7075 ISSN 2196-7083 (electronic)
ISBN 978-3-319-06307-2 ISBN 978-3-319-06308-9 (eBook)
DOI 10.1007/978-3-319-06308-9
Springer Cham Heidelberg New York Dordrecht London

Library of Congress Control Number: 2014943918

Printed on acid-free paper

Springer is part of Springer Science+Business Media (www.springer.com)

Contents

Chapter 1
Introduction

Abstract The introduction presents the main objectives and themes that have been developed in the book. The manuscript aims at offering a contribution to the ongoing debate about sustainable development in tourism, in particular by studying two phenomena able to support it: responsible tourism and Corporate Social Responsibility (CSR). After having clarified the definitions of responsible tourism, of related forms of tourism and of CSR provided by the international community, the book concentrates on the analysis of reporting systems for assessing the responsibility of tourism companies. This research effort is based on the belief that reporting programmes represent an important tool for encouraging responsible behaviour by both tourism businesses and tourists. The study develops a qualitative and quantitative analysis of reporting systems available at European level, measures their effectiveness and discusses the main issues met by companies in implementing them.

In tourism, as in other economic sectors, sustainable development represents a topic which generates a great interest and debate in the international community. Promoting growth that respects culture and environment and supports the well-being of local communities is a crucial issue for the evolution of modern tourism. This inevitably implies the increase of awareness and the adoption of responsible behaviour both by tourists and by tourism businesses involved in travel production and distribution.

This book, based on a research project carried out by CISET (the International Centre of Studies on Tourism Economics of Ca' Foscari University Venice), aims at offering a contribution to the ongoing debate, discussing two phenomena able to encourage sustainable development in tourism: responsible tourism and Corporate Social Responsibility (CSR). In particular, the attention has been paid to the study of the main reporting and certification programmes for assessing responsible tourism and CSR in the tourism industry that are available in Europe. These systems

M. Manente et al., *Responsible Tourism and CSR*, CSR, Sustainability, Ethics & Governance, DOI 10.1007/978-3-319-06308-9_1,

play an important role in driving tourism companies and SMEs to take responsibility for their own actions, adopting a more responsible business model. In addition, they enable tourists and the community to recognise and distinguish responsible tourism businesses and products and to make more conscious choices.

The book has a dual purpose: firstly, to promote a deep understanding of the most acknowledged reporting and certification systems, to identify their strengths and weaknesses and to assess their effectiveness and credibility in monitoring responsible tourism and CSR; secondly, to raise awareness on the main issues that hinder the adoption of these systems by tourism companies, in particular by small and medium-sized enterprises (SMEs), in order to suggest some possible solutions. The results aim at providing tourism businesses with a set of guidelines useful to select the "best" approach, according to their characteristics and objectives, but also to make the tourism community at large aware about the opportunity to support the implementation of reporting systems as a tool for encouraging responsible behaviour and enhancing consumers' trust.

Given these objectives, the book develops into four main parts.

- The review of literature about the different definitions of responsible tourism, of other tourism-related concepts (e.g. ecotourism, fair-trade tourism, sustainable tourism, supportive tourism, etc.) and of CSR, in order to understand and clarify the context in which reporting systems are developed (Chaps. 2 and 3).
- The analysis of the most acknowledged reporting programmes available in Europe, to raise awareness of responsible tourism and CSR in the tourism industry and to assess and certify the company responsibility. A comparative analysis has been carried out in order to identify different and common procedures/indicators adopted by each system to monitor the ecological, social and economic dimensions of sustainability and responsibility (Chaps. 4 and 5). In addition, three business case studies have been described in order to show how companies concretely adopt responsible tourism values and CSR in their business (Chap. 6).
- The assessment of the reporting systems identified according to four main criteria and through the application of a mathematic model, in order to clearly identify and summarise their main attributes and to define their effectiveness and quality in relation to SMEs' characteristics and business goals (Chap. 7).
- A final discussion on the opportunity to further promote the credibility and acknowledgement of reporting systems in the tourism community. This part discusses the implications of the study carried out and future research directions that can be followed about these themes.

Chapter 2
Responsible Tourism

Abstract Since the 1980s, the international community has become aware of the negative effects produced by mass tourism and has started to introduce the concept of sustainable development in tourism too. Forms of tourism that are considered alternative to mass tourism, above all responsible tourism, have started to emerge and nowadays their development is considered an important contribution to the sustainability of tourism practices. Given the presence of several definitions and overlapping concepts, this chapter clarifies the meaning and implications of responsible tourism and its relationship with other forms of tourism, such as ecotourism, community-based tourism, fair-trade tourism, solidarity tourism. Although each form of tourism emphasises more one specific aspect than another (environmental preservation, fairness, support to local community, etc.), all of them can be considered as a responsible way of travelling. Therefore, responsible tourism includes all those forms of travelling that pay attention to the environmental (preservation and safeguard of natural resources), socio-cultural (central role of the host community in decisions regarding local tourism development and planning; understanding of local culture and promoting cultural exchange between tourists and local people, etc.) and economic issues (fair relationships with local providers, support to local development, etc.) produced by tourism.

2.1 Sustainable Development and Alternatives to Mass Tourism

While tourism represents a very important industry, which contributes to global GDP and creates jobs, its indiscriminate development (see, for example Akama and Kieti 2007; Kasim 2006) can generate a volume of flows higher than the destination carrying capacity (Costa et al. 2001) and then produce negative impacts on the economic, social and natural environment.

M. Manente et al., *Responsible Tourism and CSR*, CSR, Sustainability, Ethics & Governance, DOI 10.1007/978-3-319-06308-9_2,

At the beginning of the 1980s, the widespread concern of the international community about the serious effects produced by mass tourism and the growing attention paid to the topic of sustainability, led to the planning of actions aimed at developing and managing tourism in a more responsible way (Anderek et al. 2007; Del Chiappa et al. 2009).

The starting point was represented by the World Tourism Conference held in Manila in 1980 and by the report "*Our Common Future*", also known as Bruntland Report, from the United Nations World Commission on Environment and Development (1987), which defined the concept of "*sustainable development*" for the first time.

The concept of "*sustainability*" was then applied to tourism too. According to the UNWTO (United Nations' World Tourism Organisation), sustainable tourism development can be defined as the tourism that "meets the needs of the present tourists and host regions while protecting and enhancing the opportunity for the future. It is envisaged as leading to management of all resources in such a way that economic, social and aesthetic needs can be fulfilled, while maintaining cultural integrity, essential ecological processes, biological diversity and life support systems" (Tesone 2004; Responsible Travel Handbook 2006).

In this sense, sustainable development in tourism emphasizes the need to preserve the traditions and the socio-cultural identity of the local community (Altinay and Hussain 2005), to protect the environment, to generate revenue and labour for host destinations (Medina 2005; Mortara 2007) and finally to involve the local community in decision-making processes of the tourism industry (Medina 2005).

At the same time of the promotion of sustainable tourism development, the concept of responsible tourism started to emerge. Understanding the meaning of responsible tourism is not an easy task, given the presence of different definitions and the proliferation of many correlated concepts in the public domain—sustainable tourism, ecotourism, fair-trade tourism, rural tourism, community-based tourism, pro-poor tourism, etc.— which are often defined as similar to responsible tourism (Bohdanowicz 2006; Responsible Travel Handbook 2006).

The concept of responsible tourism has changed over the last years. From being defined as a form of tourism characterising developing countries or destinations at the early stage of their life cycle (Clifton and Benson 2006), it has become a new way to conceive holidays, a sort of new lifestyle (Budeanu 2007) pushing a growing number of people (Meric and Hunt 1998) to make their travel choices according to values like consciousness, sobriety (Lori and Volpi 2007), responsible consumption and respect for the cultural, social, environmental and economic context of the local community. This is independently from the fact that the tourism destination chosen is domestic or international, mature or at the early stage of its life cycle (Sambri and Pegan 2007; Franch et al. 2008).

According to the Responsible Travel Handbook (2006), "responsible tourism goes beyond fancy packaging and eco-certification. It also goes far beyond simplistic internal hotel policies of washing sheets and towels, or accommodations simply being located in natural jungle or forest areas. Responsible tourism has to do with an everyday lifestyle that promotes cultural and biological diversity, and promotes environmental and natural resources conservation, at home and while travelling" (Responsible Travel Handbook 2006, p. 13). This last aspect is particularly interesting, because it underlines that people should take a responsible behaviour not only during the holiday but also when they buy and consume everyday products. From this point of view, some studies in the literature have recognised a sort of overlap between the responsible tourist and the responsible consumer in general (Sambri and Pegan 2007).

In order to understand the meaning of responsible tourism and identify its relationships with other related forms of tourism, the following section proposes and discusses the main definitions of these concepts available in the literature.

2.2 Definitions of Responsible Tourism and of Other Related Forms of Tourism

2.2.1 Responsible Tourism

According to the *Cape Town Declaration* (2002)—promoted during the first International Conference on Responsible Tourism in Destinations, held in Cape Town—responsible tourism takes a variety of forms but "it always identifies travels that:

- Minimise negative environmental, social and cultural impacts;
- Generate greater economic benefits for local people and enhances the well being of host communities, by improving working conditions and access to the industry;
- Involve local people in decisions that affect their lives and life chances;
- Make positive contributions to the conservation of natural and cultural heritage and to the maintenance of the world's diversity;
- Provide more enjoyable experiences for tourists through more meaningful connections with local people, and a greater understanding of local cultural and environmental issues;
- Provide access for physically challenged people;
- Are culturally sensitive and engender respect between tourists and hosts".

The second *International Conference on Responsible Tourism in Destinations, held in Kerala in 2008*, reviewed progress made since Cape Town in 2002 and reflected the experience gained around the world. Not only it confirmed the previous definition of responsible tourism, but also it developed some useful guidelines and recommendations for actions in different fields (education and training, campaigning and awareness raising, media, governance, partnerships, markets, etc.) in order to promote it. For example, it argued that "transparent and auditable reporting on key local social, economic and environmental issues through locally agreed indicators is considered crucial in order to enable individual consumers and businesses to make informed choices". The conference added that in determining who are the responsible tourists and what are the responsible forms of tourism it is essential to rely more on the measurement of impacts rather than to the self-declared motivations of the travellers or the companies concerned.

According to *Harold Goodwin,* director of the *International Centre for Responsible Tourism at the Leeds Metropolitan University*, "responsible tourism is about making better places for people to live and better places for people to visit" (http://www.haroldgoodwin.info/rt.html).

The *International Coalition for Responsible Tourism* associates responsible tourism with the concept of sustainability, qualifying responsible tourism as an application of sustainable development within the tourism sector. It is then defined as "any form of development or tourist activity which respects and preserves in the long term natural, cultural and social resources and contributes in a positive and fair way to the development and the bloom of people who lives, works and spend their holiday in this place" (http://www.coalition-tourisme-responsable.org/ENGLISH/tourisme-responsable.html). Therefore, according to this definition, responsible tourism should produce benefits for all stakeholders involved: the resident population, local tourism enterprises, external intermediaries (outgoing travel agencies and tour operators) and tourists. This important consideration is based on the fact that in many developing countries traditional tourists bring benefits to foreign hotel chains and tour operators only. As the Kerala Declaration recognises, they are stronger and more powerful than small and medium local businesses. In addition, local people are usually employed in underpaid jobs. Therefore, the local community cannot count on a fair distribution of profits and benefits coming from the tourism activity.

AITR (the Italian Association for Responsible Tourism) identifies responsible tourism as "tourism developed according to the principles of economic and social justice and respecting the environment and local culture. Responsible tourism acknowledges the centrality of the host community and its right to be the main driver of the sustainable and socially responsible development of their territory. This form of tourism acts supporting the positive interaction between the tourism industry, the local community and travellers" (http://www.aitr.org). According to the AITR perspective, the focus is on social aspects, such as the participation of the local community to decisions and a fair exchange among all stakeholders involved.

Likely, in *Spain* responsible tourism has been defined as "the tourism product or the system of tourism production in which the tourists and the local stakeholders take responsibility over the destination from the environmental, social, cultural and economic point of view, recognizing the central role of the local community" (Palomo Perez 2006; http://www.turismoresponsablehoy.org/). This definition seems to be based on the same values declared by AITR, with the difference that the Italian organisation also emphasises the importance of an advantageous interaction between tourists, local communities and tourism industry.

A new contribution on responsible tourism definition also comes from the *European Charter for a Sustainable and Responsible Tourism*, a document developed by the European Union which is now being discussed by the community (http://ec.europa.eu/enterprise/sectors/tourism/sustainabletourism/charter/documents_charter/summary_consultation_en.pdf). Aimed at representing a common reference point for all stakeholders in tourism and defining nine basic principles that encourage the development and implementation of responsible policies and behaviour, the Charter clarifies the concept of responsible tourism. It "stands for responsibility for decisions, actions and policies of all those involved in the planning, management, delivery and consumption of tourism, so that it is sustainable over time". The charter also specifies that responsible tourism makes possible a sustainable development of tourism.

It is evident that different organisations and subjects have proposed their own definition of responsible tourism, partly sharing the same values and partly promoting a specific perspective. In addition, definitions tend to change according to the country where responsible tourism develops or where responsible tourists originate from.

2.2.2 *Forms of Tourism Related to Responsible Tourism*

As mentioned before, not only the lack of a universally accepted and shared definition affects responsible tourism, but also other correlated forms of tourism, which are characterised by an overlapping of concepts and contributions.

A widespread and often misused form of tourism is *ecotourism* (Franch et al. 2008). According to the definition given by the World Ecotourism Summit organised by UNWTO in 2002, "ecotourism is used to mean all nature-based forms of tourism (http://sdt.unwto.org/en/content/ecotourism-and-protected-areas):

- In which the main motivation of the tourists is the observation and appreciation of nature as well as the traditional cultures prevailing in natural areas;
- Which contains educational and interpretation features;
- Which are generally, but not exclusively organised by specialised tour operators for small groups. Service provider partners at the destinations tend to be small, locally owned businesses;
- Which minimise negative impacts upon the natural and socio-cultural environment;
- Which support the maintenance of natural areas which are used as ecotourism attractions by:
  - Generating economic benefits for host communities, organisations and authorities managing natural areas with conservation purposes;
  - Providing alternative employment and income opportunities for local communities;
  - Increasing awareness towards the conservation of natural and cultural assets, both among locals and tourists".

This definition can be integrated by the contribution given by Planeta.com (Responsible Travel Handbook 2006), which argues that ecotourism should meet three criteria:

- It provides for environmental conservation;
- It includes meaningful community participation;
- It is profitable and can sustain itself.

In addition, according to TIES (The International Ecotourism Society), ecotourism identifies "responsible travels to natural areas that conserve the environment and improve the well-being of local people" (Responsible Travel Handbook 2006).

These three contributions, which represent the best known definitions of ecotourism, share the same principles (the focus on the environmental and cultural preservation and on the profitability for the local community), even though they express the concept in a different way. In particular, TIES' definition seems to consider ecotourism as a form of responsible tourism.

Another form of tourism that is related to responsible tourism is *fair-trade tourism*. It takes inspiration from the same values characterising fair trade, aimed at helping producers, in particular in developing countries, to make better trading conditions and promote sustainability. Fair-trade tourism brings fair trade values into travel. "This means working with the travel industry to make things fairer for people living in what are traditionally known as "destinations". Fair-trade tourism is guiding the way toward sharing benefits more equitably between travellers, the tourism industry, governments of the countries visited, and most importantly, the host-country nationals" (Responsible Travel Handbook 2006).

More complete is the definition of *tourisme equitable* (in English, fair-trade tourism) promoted in France as "a group of tourist service activities proposed by tourist operators to responsible travellers and totally or partially developed by local communities. These communities also take part in the management of these activities. Economic, social and cultural benefits have to be spread locally and go for most part to the benefits of local population" (Tourisme Solidaire: http://www.tourismesolidaire.org/).

It is interesting to highlight that this definition, as that of ecotourism proposed by TIES, refers to the concept of responsible tourism and travel (in this case "responsible travellers"). From this point of view, also fair-trade tourism seems to be considered part of responsible tourism.

Another widespread concept—for example in France—is *solidarity tourism*, which groups all the "alternative forms of tourism that put at the core of travel the meeting with local communities and that aim at developing the local territory. The involvement of local populations in different steps of the tourism project, the respect for people, culture and nature and the equal allocation of the resources created by the project, are the basic foundations of this kind of tourism" (Tourisme Solidaire: http://www.tourismesolidaire.org/).

A further form of tourism, which is alternative to mass tourism and related to responsible tourism, is *community-based tourism*. According to the Responsible Travel Handbook (2006), it is defined as "a holistic approach to tourism that incorporates the environmental, socio-cultural, and economic impacts of tourism and that includes the basic goals of ecotourism but with a few enhancements:

- Travel to natural destinations inhabited by indigenous cultures, learning from them and directly helping the disappearing indigenous communities around the world through cultural exchange, financial assistance, and education;
- Minimize negative impacts, by encouraging and supporting environmentally sensitive practices, not only by travellers but also by local people;
- Build awareness, promoting the exchange of knowledge and wisdom for both visitors and residents of host communities;
- Provide financial benefits and empowerment to indigenous people, by helping them to maintain their right to self-determination, giving them decision-making authority regarding the conduct of tourism in their lands;
- Respect local culture, since sensitivity shouldn't stop with the ecosystem but extend to understanding and respecting cultures in their own context".

Whether this overview about those forms of tourism alternative to mass tourism and related to responsible tourism seems very comprehensive and varied, the list can be even longer, if other tourism practices that share some of the principles of responsible tourism have been taken into account (Responsible Travel Handbook 2006):

- *Rural tourism*: travels in non-urban destinations, participating in or observing activities, events, or attractions that are a fundamental part of rural communities and environment;
- *Agro-tourism*: a subcategory of ecotourism and rural tourism, which encourages visitors to experience and learn about agricultural life for periods of a day, overnight or longer-term, also working in the fields alongside farmers, coffee growers, vineyardists or fishermen;
- *Conscientious tourism*: travels that encourage a deeper interaction and understanding of people and place, also by engaging travellers in various activities;
- *Heritage tourism*: tourism that respects natural and built environments, in short the heritage of the people and place, also contributing to local economy and well-being;
- *Reality tourism:* socially responsible travel founded on the principles of experiential education and on important social, economic, political and environmental issues of the destination; the emphasis is on meeting the people, learning the facts firsthand and then working toward the alleviation of global problems and enacting positive change;
- *Pro-poor tourism*: not a specific product or niche sector but an approach to tourism development and management, enhancing the linkages between tourism businesses and poor people, so that tourism's contribution to poverty reduction is increased and poor people are able to participate more effectively in product development.

It is finally interesting to underline that the Responsible Travel Handbook (2006), which is a guide for increasing awareness on responsible tourism, does not give any specific definition of responsible tourism. Rather it clarifies that responsible tourism includes many challenges and alternatives to mass tourism and describes different forms of niche tourism—those mentioned just above—thus arguing that all of them are an expression of responsible tourism.

The same understanding seems to be supported by Laurent (2003), who states that responsible tourism is a sort of umbrella term, which includes all type of tourism mentioned just above, whose borders are fuzzy and dynamic: ecotourism, rural tourism, solidarity tourism, fair-trade tourism, community-based tourism, etc.

2.3 Comparing Concepts and Definitions: Common Aspects

As Del Chiappa states (2009), differentiating the concept of responsible tourism from that of other forms of tourism is not an easy task. Firstly because a high number of definitions were given over the time and with similar and ambiguous meanings (Kasim 2006). Secondly, because of the proliferation of correlated and overlapping terms and of the principles behind them, i.e. their ethical implications and then the actions and behaviour taken by all stakeholders involved.

However, the overview proposed in the previous paragraph shows that, although each form of tourism emphasizes more one specific aspect rather than another, all of them share the same principles of responsible tourism. For example, ecotourism focuses on environmental preservation, but it also highlights the value of respect, fairness and equity towards the local community. Fair-trade tourism states that benefits from tourism have to be shared more equitably between the tourism industry and the host community in the destination, but it also takes into account other values of responsible tourism. Finally, solidarity tourism seems to integrate the principles of responsible tourism with those of fair-trade tourism.

Therefore, trying to summarise the *key elements* and recommended behaviour included in all definitions analysed, it is possible to identify six main requirements that address all tourism stakeholders involved.

- The host community should participate in the decision-making process and in the local planning of tourism development;
- Tourism activity should be a source of income and jobs for the host community, contributing to economic growth and well-being of people in the destination;
- Benefits and revenue from tourism should be fairly distributed among local operators;
- Nature, environment and culture should be respected and preserved;
- Travels should promote meaningful interactions, cultural exchange and understanding of local people.

Table 2.1 compares the main elements characterising responsible tourism and other alternative forms of tourism, showing the relevance/intensity of each element

Table 2.1 Responsible tourism and related forms of tourism: relevance of economic, social and environmental aspects

	Community involvement	Local development	Equity	Environmental preservation	Respect for different cultures	Meeting with local population
Responsible tourism	5	4	2	3	3	3
Ecotourism	3	3	1	5	2	2
Fair-trade tourism	4	3	5	1	1	1
Solidarity tourism	5	4	2	3	3	4
Community-based tourism	5	4	2	3	3	2

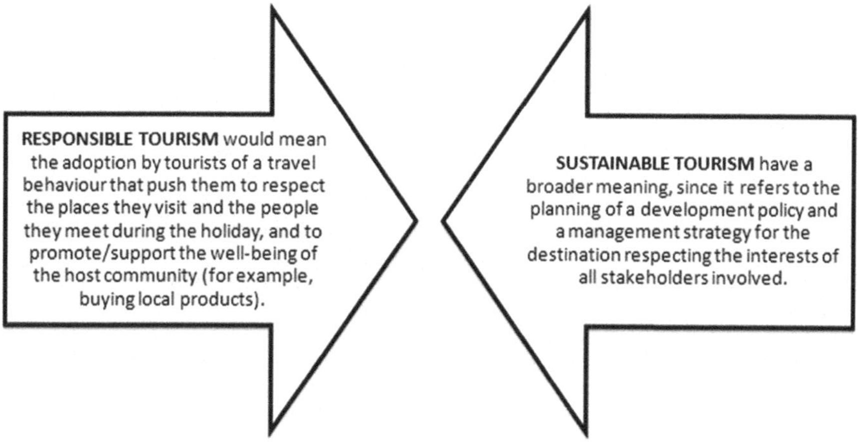

Fig. 2.1 Difference between responsible tourism and sustainable tourism (Authors' elaboration from Colombo 2005; Sambri and Pegan 2007)

analysed. The relevance is measured using a scale (1–5) with 1 = minimum relevance and 5 = maximum relevance.

Generally speaking, if developed respecting the elements highlighted in Table 2.1, all the alternative forms of tourism can be considered as a responsible way of travelling, respecting and supporting cultures, environment and local communities. Therefore they can be included in a broad definition of responsible tourism. The basic elements are the same, but their importance changes in intensity according to the type of tourism taken into account.

In addition, it is useful to clarify the relation between responsible tourism and sustainable tourism. Recently, some new contributions in the literature have suggested that the two concepts entail a different perspective of definition. As explained in Fig. 2.1, responsible tourism would imply a demand-side perspective and sustainable tourism a supply-side one (Colombo 2005; Sambri and Pegan 2007).

In other words, responsible tourism is a broad concept that includes different conscious and respectful ways of travelling, which bring people to take responsibility for their actions. Sustainable tourism represents the natural reply that tourist operators give in order to fulfil the needs and wishes of responsible tourists.

References

Akama JS, Kieti D (2007) Tourism and socioeconomic development in developing countries: a case study of Mombasa Resort in Kenya. J Sustain Tourism 15(6):735–748

Altinay M, Hussain K (2005) Sustainable tourism development: a case study of North Cyprus. Int J Contemp Hospit Manag 17(3):272–280

Anderek KL, Valentine KM, Vogt CA et al (2007) A cross-cultural analysis of tourism and quality of life perceptions. J Sustain Tourism 5(15):483–502

Bohdanowicz P (2006) Environmental awareness and initiatives in the Swedish and Polish hotel industries – survey results. Hospit Manage 25(4):662–682

Budeanu A (2007) Sustainable tourist behaviour – a discussion of opportunities for change. Int J Consumer Stud 31:499–508

Cape Town Declaration on Responsible Tourism in Destinations (2002) http://responsibletourismpartnership.org/CapeTown.html. Accessed 16 July 2013

Clifton J, Benson A (2006) Planning for sustainable ecotourism: the case for research ecotourism in developing country destinations. J Sustain Tourism 14(3):238–254

Colombo L (2005) Il turismo responsabile. Xenia Edizion, Milano

Costa P, Manente M, Furlan MC (2001) Politica economica del turismo. TUP, Touring Editore, Milano

Del Chiappa G, Grappi S, Romani S (2009) The responsible tourist's behaviour: an empirical analysis in Italy. In: Paper presented at the 8th international congress marketing trends, Paris, 16–17 Gen 2009

Franch M, Sambri C, Martini U et al (2008) La domanda di turismo responsabile e di ecoturismo in Italia. Un'indagine esplorativa sui turisti CTS. In: Paper presented at the 7th international congress marketing trends, Venice, 17–19 Jan 2008

Kasim A (2006) The need of business environmental and social responsibility in the tourism industry. Int J Hospit Tourism Admin 7(1):1–22

Laurent A (2003) Caractériser le tourisme responsable facteur de développement durable. Direction Generale de la Coopération Internationale et du Développement. Ministères des Affaires Etrangère, France

Lori M, Volpi F (2007) Scegliere il "bene". Indagine sul consumo responsabile, FrancoAngeli, Milano

Medina LK (2005) Ecotourism and certification: confronting the principles and pragmatics of socially responsible tourism. J Sustain Tourism 13(3):281–295

Meric HJ, Hunt J (1998) Ecotourists' motivational and demographic characteristics: a case of North Carolina travelers. J Travel Res 36(4):57–61

Mortara A (2007) Il consumatore etico esiste davvero. In: Paper presented at the 6th International Congress Marketing Trends, Venice, 26–27 Jan 2007

Palomo Perez S (2006) El turismo justo y la creación de sistemas de producción de servicios turísticos responsables. Estudios Turísticos 168:7–46

Responsible Travel Handbook (2006). http://www.transitionsabroad.com. Accessed 12 March 2010

Tesone DV (2004) Development of sustainable tourism hospitality human resources management module: a template for teaching sustainability across the curriculum. Hospit Manage 23:207–237

Sambri C, Pegan G (2007) La domanda di turismo responsabile: risultati di un'indagine esplorativa in Italia. In: Paper presented at the 6th International Congress Marketing Trends, Venice, 26–27 Jan 2007

United Nations – World Commission on Environment and Development (1987) Our common future. http://www.un-documents.net/wced-ocf.htm. Accessed 25 Aug 2013

Chapter 3
Corporate Social Responsibility in the Tourism Industry

Abstract Since the mid-1990s, the concept of Corporate Social Responsibility (CSR) started to emerge and be applied in different industry sectors. CSR arises from several issues: for example the businesses' need to gain and retain consumer trust and the awareness that companies should take their responsibility for the impacts their activities cause on environment and society. CSR is therefore based on the concept of "triple bottom line", which replaces the financial bottom line and implies that businesses are responsible for the environmental, social and economic effects they produce. This chapter clarifies the meaning and implication of CSR; explains the role of CSR in building and retaining consumer trust and, finally, explores the state of the art of CSR diffusion in tourism, highlighting the main benefits it brings to the tourism industry and the issues met by tourism companies, in particular by small- and medium-sized enterprises. The analysis points out that although CSR has a positive influence on consumer trust, efficiency, tourism product quality and business competitiveness, its implementation in the tourism industry, and especially in tourism SMEs, is limited by lack of awareness and knowledge and by high investments needed to implement an appropriate action plan.

3.1 Definitions of CSR

The concept of Corporate Social Responsibility (CSR) has been developing since the mid-1990s outside the tourism sector, becoming an essential framework for changing management practice in general. Supporters of CSR, such as Kalish (2002), argue that CSR can contribute to increase consumer confidence in a context where customers and society at large have decreased their trust in business standards and ethical behaviour. In addition, major industry players see the principles and practices of CSR as a tool and a pre-requisite for responsible and sustainable

M. Manente et al., *Responsible Tourism and CSR*, CSR, Sustainability, Ethics & Governance, DOI 10.1007/978-3-319-06308-9_3,

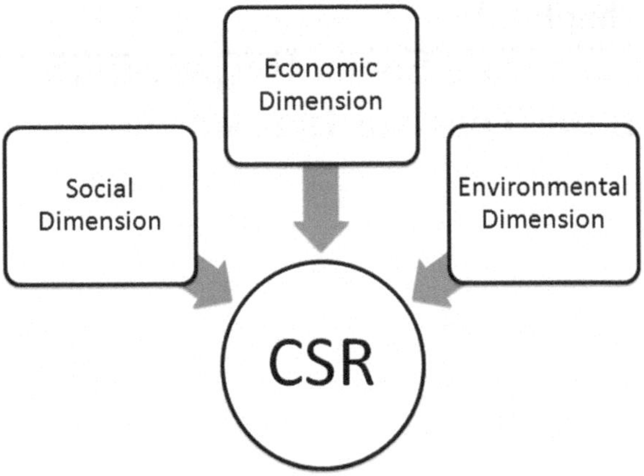

Fig. 3.1 Triple bottom line (Authors' elaboration from WBCSD 1999)

development of industrial and service operations. From this point view, CSR is based on the premise that sustainability cannot be achieved if businesses do not take responsibility for their actions towards society.

Therefore, while in the past the financial bottom line only was acknowledged, now it is common to use the term of *triple bottom line*, referring to the fact that businesses have to be aware and responsible for the economic, social and environmental impacts they produce with their activities (Fig. 3.1). Beyond maximising profits, they have to act respecting the staff, the local community, the environment and the other stakeholders (Kalish 2002).

However, CSR lacks in a generally accepted definition. Several approaches have been proposed over the years by different schools of managerial thought. "For some, it is simply a management approach in which the interests of all the organisation's stakeholders are identified, assessed and respected. For others, CSR is a dynamic process of recognizing and responding to evolving forms of accountability between the organisation and its stakeholders" (White 2006).

According to the *Business for Social Responsibility*, CSR means "achieving commercial success in ways that honour ethical values and respect people, communities and the natural environment" (White 2006). This definition stresses the idea of recognizing and responding to a broader spectrum of stakeholder interests that includes, but it is not limited to, shareholders.

The *World Business Council for Sustainable Development* (WBCSD 1999) acknowledges that CSR is "the continuing commitment by business to behave ethically and contribute to economic development while improving the quality of life of the workforce and their families as well as of the local community and society at large". It is then driven by a commitment to human rights, employees rights, environmental protection, community involvement, supplier relations, stakeholder rights and monitoring (http://www.wbcsd.org/work-program/business-role/previous-work/corporate-social-responsibility.aspx).

A similar definition is given by the World Bank, who claims that CRS describes the commitment of a company to be responsible toward stakeholders in all its activities. A socially responsible company considers the impacts its produces on the environment and the local community when it takes decisions, balancing the interests of all stakeholders while making profits (World Bank 2003).

The European Commission too has recognized the role of CSR as a contributing factor to the Lisbon Strategy for growth and jobs, in that it can help to shape the kind of competitiveness model that Europe wants. According to the definition coined in 2001, "being social responsible means not only fulfilling legal expectations, but also going beyond compliance and investing more into the human capital, the environment and the relations with stakeholders" (EC COM (2001) 366, p. 6). CSR is then defined by "a concept whereby companies integrate social and environmental concerns in their business operations and in their interaction with their stakeholders on a voluntary basis". This means that:

- CSR covers economic, social and environmental issues at the same time, in spite of the English term "Corporate Social Responsibility";
- CSR is not or should not be separate from business strategy and operations. It is about integrating social and environmental concerns into the enterprise's activities;
- CSR is a voluntary concept;
- An important aspect of CSR is how enterprises interact with internal and external stakeholders (employees, customers, local communities, neighbours, non-governmental organizations, public authorities, etc.).

In 2006, the Commission published a new Communication (EC COM (2006) 136 final) emphasising the importance of CSR and challenging businesses to take leadership in this field. According to this document, CSR can play a key role in contributing to sustainable development and to the competitiveness of enterprises of all sizes. In particular, the Commission aims at making Europe a pole of excellence of CSR, also launching the European Alliance for CSR, a political umbrella for new or existing CSR initiatives. The main goals of this initiative were:

- Raising awareness and improving the knowledge on CSR and reporting on its achievements;
- Helping to mainstream and develop open coalitions of co-operation;
- Ensuring an enabling environment for CSR.

Following these statements and further stressing that not only CSR is in the interest of society as a whole but also of companies, in 2011 the European Commission published a new document, proposing a revised and enlarged definition of CSR, which is now conceived as "the responsibility of enterprises for their impacts on society" and incorporates all principles mentioned in the previous one (EC COM (2011) 681 final). In this way, the Commission clarifies that, for fully meeting their corporate social responsibility, enterprises should have in place a process to integrate social, environmental, ethical, human rights and consumer concerns into their business operations and core strategy in close collaboration with their stakeholders.

Table 3.1 Eight main areas of the CSR principles (Gordon 2001)

• Accountability
Stakeholder engagement, reporting on environmental and human rights issues, performance related to standards.
• Business conduct
Competitive conduct, corruption and bribery, proprietary information, intellectual property rights, political activities.
• Community involvement
Community economic development, employment of local and/or under-utilised workers, philanthropy.
• Corporate governance
Right of shareholders, conduct of executive boards
• Environment
Precautionary principle, input/output, stakeholder engagement, training of employees, management systems, public policy, sustainable development.
• Human rights
Indigenous peoples' rights, health and safety, child/forced labour, freedom of association, wages and benefits, working conditions, discipline.
• Marketplace/consumers
Marketing/advertising, product quality and/or safety, consumer privacy.
• Workplace/employees
Non-discrimination, training, harassment/abuse, downsizing, child/elder care, maternity/paternity leave.

Considering the main definitions of CSR and according to the principles developed by different organisations, eight core areas of CSR can be identified within a company, from human resources management to suppliers' selection, from production to relationship with consumers, competitors, local stakeholders, etc. (Gordon 2001) (Table 3.1).

CSR should then be integrated into every aspect of business strategy, policies, processes and practices, trying to involve all stakeholders and players along the *value chain*. Therefore, this approach goes beyond the initial perspective, which conceives CSR only as a philanthropic action or the maximisation of shareholders' benefit (Kalish 2002).

However, CSR is often implemented by companies in an incomplete way, being a *melting pot* of actions that are uncoordinated, not consistent with the core business and not integrated in the internal policies and strategies. The spectrum of activities ranges, for example, from sponsorship of cultural and sport events to planting trees for climate protection, from donations and charity galas to projects for poor or ill children. Even though these initiatives are commendable, they are more linked to the company's corporate citizenship than to a real CSR (Ermlich 2009).

3.2 The Role of CSR in Building and Retaining Consumer Trust

Products based on ethical, social and environmental values or developed by responsible businesses will primarily attract consumers who are interested in specific issues or who are particularly receptive to these aspects of purchasing (Vitell et al. 2003). Responsible consumers—in this case responsible tourists—are demanding and cautious in their choices, since they want to buy products that are truly conceived in full compliance with ethical and environmental values (Goodwin 2005).

In this process, they frequently encounter difficulties that can damage their confidence in businesses and products. Firstly, it is hard for consumers to distinguish authentic responsible products in the market, since they do not have the opportunity to assess whether the responsible values claimed by the company are respected or not: this also considering that "responsibility" is not an attribute that consumers can easily identify and check in the product (Gebben and Gitsham 2007). Secondly, consumers are concerned that companies and final distributors keep for themselves most of the extra-charge- they are generally required to pay for responsible products in comparison to others in order to finance social or environmental projects and fairly pay local providers (Harford 2005; Weber 2007). There is therefore an information asymmetry that prevents consumers from completely trusting the company's reliability to respect the promise that implicitly characterises responsible products.

It is evident that all these issues impact on consumer trust. This concept can be defined as the consumer's expectation that the business is willing to keep promises and fulfil obligations. Consumer's expectation is based on company's competencies, honesty, altruism and goodwill and the higher the level of these values, the greater the expectation (Castaldo et al. 2008). Trust is considered particularly crucial when there is a perceived risk and no control systems are available (Schoorman et al. 2007) or in business relationships where the person who trust (in this case the consumer) is in a vulnerable position (Pivato et al. 2008). Starting from all these considerations, it is a foregone conclusion to assume that responsible consumers are in a vulnerable position and that trust is a core element in their intention to buy responsible products.

Empirical studies in different sectors—as organic food and fair trade products—have found that consumer trust in responsible goods is positively and strongly correlated with consumer perception of business' CSR performance and with consumer trust towards private labels, certifications and reporting systems (Castaldo et al. 2008; Pivato et al. 2008). Consumer perception that a company is socially and environmentally oriented is associated with a higher level of trust in that company and its products. Indeed consumers are inclined to believe that a reputable company, which has based its business on a positive relationship with the community and on environmental preservation, has strong incentives to maintain its reputation and is less likely to renege on promises. Trust, in turn, influences consumers' subsequent actions and then the intention to purchase a responsible product (Pivato et al. 2008).

However, the link seems to exist when at least two conditions prevail: the products sold by the company seriously comply with ethical, social and environmental requirements, and the company itself has an acknowledged commitment (Castaldo et al. 2008). Because of that, social and ethical accountability, sustainability reports, codes of conduct, reporting systems and labels are important sources of information that support consumers and the community at whole to check the company. With the help of these tools, consumers recognise a business following CSR as better than other competitors that are not CSR oriented (Pivato et al. 2008).

Therefore, although research on the relationship between CSR and consumer trust should be further improved, studies available in the literature seem to demonstrate that the more a company is committed towards CSR and is able to appropriately disseminate evidences about its efforts, the more customers are likely to trust and reward it, by buying its products (Fig. 3.2).

Studies carried out in fields related to responsible tourism, such as fair trade products and organic food, suggest that the positive relationships between CSR and consumer trust is valid in tourism too. Therefore, the adoption and implementation of CSR by tourism companies prove their commitment towards a responsible path and stimulate tourists' confidence and willingness to buy responsible packages and holidays. For these companies and products CSR can be a source of competitive advantage, because competitors with bad or social and environmental reputation will be unable to build and retain consumer trust.

In addition, it can be stated that not only trust is strategic for responsible travellers but also for tourists in general. As any other service, tourism product is intangible and experienced based, i.e. it is founded on a promise the company makes to the traveller and its attributes cannot be verified before consumption. Therefore the purchase of a holiday is almost a speculative investment, involving a high degree of risk and then of trust (Mitchell and Vassos 1997; Holloway 2002).

There is reason to argue that tourists are likely to consider companies committed to CSR to be more reliable than non CSR-oriented operators and to trust them and buy their products. This because not only tourism businesses that have implemented CSR are committed to protect the environment and support local community but also to respect consumer's rights, by ensuring tourists' health and safety during the holiday and the quality of the product.

CSR seems to support a favourable context that positively enhances customers' perception and attitude towards the firm. Tourists who can rely on the existence of CSR are likely to reduce their uncertainty (Nicolaus 2008). Also research carried out on products that not necessarily imply a specific responsible consumption—such as cosmetics and sportswear— show that consumers' perceptions of CSR activities positively impacts on their trust towards the company, through the influence on perceived quality of the product (Swaen and Chumpitaz 2008).

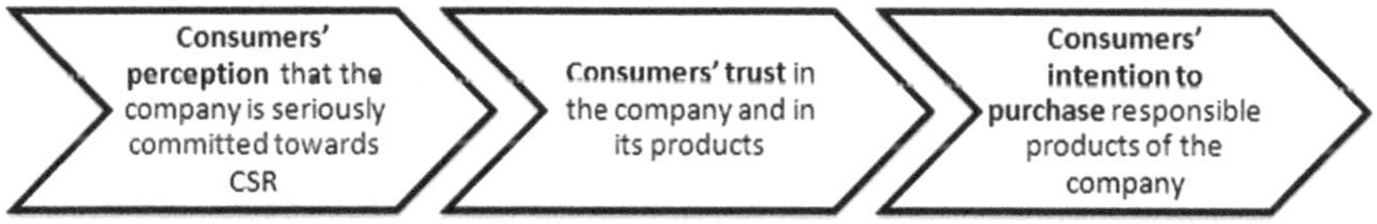

Fig. 3.2 Relationship between CSR and consumer trust (Authors' elaboration from Pivato et al. 2008)

3.3 CSR in the Tourism Industry: Main Issues

As a service industry using people and environment at national and international level, tourism should bear major responsibility for sustainable development and respect for the environment and human rights (Kalish 2002).

There is the need for tourism development "to be guided by the wishes of local people, recognising the contribution that people, customs and lifestyles make to the tourism experience, and recognising that local people must have an equitable share in the economic benefits of tourism" (Kalish 2002). In addition, tourism operators and destinations should realise the negative impacts that tourism can have on their product and be aware that the resources that attract tourists need to be protected for long-term business sustainability (Dodds and Kuehnel 2010).

This consideration means for tourism companies (Dodds and Joppe 2005):

- To adopt a long-term vision;
- To implement a triple bottom line approach and a multi-stakeholder approach;
- To protect and conserve sustainable resources;
- To be environmentally responsible;
- To be equitable;
- To maintain the well-being and involvement of local population or host;
- To provide meaningful and fairly remunerated employment for the host population;
- To have economic benefit;
- To obtain optimum guest satisfaction and educate tourists about environmental and social concerns.

Consequently, CSR has many similarities with sustainable tourism, in that both "focus on how stakeholders should be identified and engaged and what initiatives should be measured to determine their impacts on others" (Dodds and Joppe 2005, p. 9).

Although, as stated by the United Nations, "CSR is no longer a luxury for the travel and tourism sector but a need in the global marketplace" (Harms 2009 in Ermlich 2009, p. 2), it is still unexplored in comparison with other industries" (Ermlich 2009; Frey and George 2010; Coles et al. 2013; Tamajón and Font 2013). While in these industries (e.g. food, textiles, information technology, etc.), CSR standards have been developed by companies to respond to external pressure, in the tourism sector many companies do not apply codes of conduct and certification, that in turns are not based upon agreed international standards (Dodds and Joppe 2005).

This is because the characteristics of the tourism sector (high fragmentation, cross-sectional, prevalence of small and micro enterprises, etc.) and of the product (heterogeneity, intangibility, etc.) make it difficult for most enterprises to accomplish the strict requirements of these standards. In addition, while international hotel companies and big tour operators have already adopted individual ad hoc CSR initiatives, SMEs (small- and medium-sized enterprises) find it difficult to organise their core business in a socially and ecologically responsible way and to develop the conditions under which a product can be defined as "responsible" (Ermlich 2009). They often do not have appropriate knowledge of the destination, understanding and skills in community development and the budget to invest in long-term planning such as a CSR strategy requires (Miller 2001; Tamajon and Font 2013). This issue is particularly serious since SMEs represent the majority of tourism industry: in Europe, for example, they accounts for about 90 % of total businesses.

Another important aspect which limits the adoption of CSR by tourism companies is the high costs of reporting and certification systems assessing the business responsibility. Many programs are so costly to run, time-consuming and labour-intensive that they marginalise small businesses. Together with subscription fees and the expenses for consultancy and independent auditing, there are the costs for the reconfiguration of internal processes (Dodds and Joppe 2005; Rome 2007; Bohdanowicz-Godfrey 2007; Russillo et al. 2007; Jarvis et al. 2010; Bobbin 2012a). In particular, it is possible to distinguish between direct and indirect costs. Direct costs are generally limited to the short term and directly linked to the implementation of the reporting system: they include application and membership fees, inspection and visit costs. Indirect costs instead refer to the long term and include investments for implementing changes, rethinking processes and financing the equipment and infrastructure necessary to meet requirements. For example, the cost of moving from free disposal of sewage to a system of treating sewage or changing from low capitol cost equipment such as a petrol generator to a solar-powered system (Russillo et al. 2007) (Table 3.2).

Whether in large companies the costs of implementing CSR are affordable, given their budget and expertise, for SMEs the process is more complicated to be managed and supported. In addition, also for big and medium tourism companies, CSR efforts seem to not have produced the promised marketing benefits, as increase of consumer awareness (Dodds and Joppe 2005). A study commissioned by the World Bank Group has shown that, although the significant development of reporting systems over the past 10 years, SMEs still do not have greater access to market opportunity and the overall tourism industry is still not committed towards responsibility (Dodds and Joppe 2005). Aware of this issue and of the difficulties met by SMEs, some organisations have started to develop different solutions to enable small businesses to implement their own reporting system: less expensive fees, free assistance and advice, support in finding appropriate sources of funding, development of networks, exchange of best-practices, easier processes and standards, etc. (Rome 2007).

Table 3.2 Reporting programs: direct and indirect costs (Russillo et al. 2007)

Direct costs
Fees for the adhesion to the reporting system
Tecnical Documents
Logo exhibition
Auditing
Indirect costs
Investments for respecting minimum requirements:
New technologies
New infrastructure
New management system
Social and environmental programs
Quality standards
Training courses for the staff

A last aspect to be taken into account is the plethora of the available certification programs, most of which are not acknowledged at international level and appropriately promoted. The increasing demand for responsible tourism has encouraged the industry to develop reporting programs, certifications, labelling and award schemes to grade CSR businesses (Dodds and Joppe 2005; Goodwin 2005; Font and Epler-Wood 2007; Bobbin 2012b). Although variety is desirable, because it reflects the cultural and ecological diversity of people and countries (Goodwin 2005), the proliferation and little recognition of schemes can generate great resistance on the part of tourism companies and consumers, who are not able to distinguish among systems and to identify the most effective and credible program (Dodds and Joppe 2005). This explains why, despite the benefits they brought, reporting programs continue to meet a low market penetration and an apparent consumer indifference (Buckley 2013).

However, a stimulus to drive the adoption of CSR can be represented by the benefits it can bring to companies—both large and small businesses—by covering the costs met for its implementation, contributing to the business performance and producing extra revenues and marketing advantages. In this sense CSR is in the interest not only of society as a whole but also of companies.

Benefits brought by CSR can change in nature and intensity according to the type of company and the sector it belongs to and generally occur over a medium-long period of time. Together with the opportunities to increase and retain consumer trust (see Sect. 3.2), the other most important benefits are summarised in Table 3.3 (Gordon 2001; Kalish 2002; Nurn and Tan 2010; Martínez and Rodríguez del Bosque 2013).

Finally, being tourism a people-centred industry, based on the interaction of tourists with human resources of the company and with the host community in the destination (Kalish 2002), and being the environment and the cultural heritage core attractions of the tourism product (Crouch and Ritchie 2003), CSR is even more important in tourism. CSR can support tourism businesses in building profitable and collaborative relationships with employees, host communities and local suppliers and in guaranteeing a proper use of resources—that are unique and

Table 3.3 Benefit brought by CSR to companies (Gordon 2001; Kalish 2002; Nurn and Tan 2010; Martínez and Rodríguez del Bosque 2013)

Internal savings	Adopting alternative and more efficient systems for energy and water consumption, waste reduction and recycling, etc. are actions that contribute to preserve the environment and to reduce internal operating costs, without impairing the quality of services delivered
Differentiation and ability to respond to new consumers' requirements	Even though tourists still show great resistance to concretely change their purchase and consumption behaviour towards a responsible path, the number of people concerned about the environment and interested in having a true contact with the destination and the local community is growing. There is a small but non negligible segment of really responsible tourists who seriously adopt conscious and respectful choices and behaviour before and during the holiday and who require companies to produce responsible goods and services. Therefore, a business that implements CSR can be more competitive in attracting this market segment and in the long term it is likely to be rewarded with increased sales
Better risk management	Applying a CSR strategy can help the enterprise to better manage the internal and external process (e.g. production, human resources, relationships with partners, customer satisfaction, etc.) and improve its image and reputation. For example, the production of goods and services that directly or indirectly employs children' work or that has negative effects on the environment can push consumers to boycott the business. On the contrary, taking on a social and environmental responsibility can contribute to enhance the business' reputation and improve its long-term profitability
Preventing future regulations	Governments are undergoing an increasing pressure, supported by non-profit organisations and the public community at large, to strengthen the controls on business's activity and impose new regulations about environmental preservation, ethics issues, etc. These new regulations generally translate for businesses in an extra cost in order to adapt internal processes within the time-frame imposed by the law. Companies that have already implemented and integrated CSR into their internal operations can be advantaged and in a better position than their competitors, because they have already fulfilled the requirements of the law (or they have to adopt only minimal interventions). In addition, due to their experience, they can affect policy makers' decisions taken at local, national or global level

not-reproducible—balancing the need for their preservation. Adopting this approach means investing in human resources' education, training and satisfaction, in the development of employees' know-how, competences and willingness to provide high-quality services, in the well-being of host communities, in the conservation and promotion of natural and cultural heritage. All these elements contribute to the quality of the tourism product and finally to customer satisfaction and loyalty (Gordon 2001; Levy and Duverger 2010; Martínez and Rodríguez del Bosque 2013; Prud'homme and Raymond 2013).

References

Bobbin J (2012a) Researching responsible tourism reporting. International Centre for Responsible Tourism, Occasional Paper OP. 24

Bobbin J (2012b) Introduction to responsible tourism reporting. The development of a transparent verifiable reporting system suitable for small to medium sized enterprises to monitor their responsible tourism good practices. Responsible Tourism Reporting. http://www.rtreporting.org/about/news-and-publications.php. Accessed 8 Aug 2013

Bohdanowicz-Godfrey P (2007) A case study of Hilton Environmental Reporting as a tool of Corporate Social Responsibility. Tourism Rev Int 11(2):115–131

Buckley R (2013) Social-benefit certification as a game. Tour Manage 37:203–209

Castaldo S, Perrini F, Misani N et al (2008) The missing link between corporate social responsibility and consumer trust: the case of fair trade products. J Bus Ethics 84:1–15

Coles T, Fenclova E, Dinan C (2013) Tourism and corporate social responsibility: a critical review and research agenda. Tourism Manage Perspect 6:122–141

Commission of the European Communities (2001) Promoting a European Framework for Corporate Social Responsibility. COM (2001) 366 final

Commission of the European Communities (2006) Communication from the Commission to the European Parliament, the Council and the European Economic and Social Committee. Implementing the Partnership for Growth and Jobs: making Europe a Pole of Excellence on Corporate Social Responsibility. COM (2006) 136 final

Commission of the European Communities (2011) A renewed EU strategy 2011–2014 for Corporate Social Responsibility. COM (2011) 681 final

Crouch GI, Ritchie JRB (2003) The competitive destination: a sustainable tourism perspective. CAB International, Oxfordshire, UK

Dodds R, Joppe M (2005) CSR in the tourism industry? The status of and potential for certification, codes of conduct and guidelines. Foreign Investment Advisory Service Investment Climate Department, IFC/World Bank, Washington

Dodds R, Kuehnel J (2010) CSR among Canadian mass tour operators: good awareness but little action. Int J Contemp Hospit Manag 22(2):221–244

Ermlich G (2009) Corporate social responsibility (CSR). ITB Berlin – Special Press Release 04/2009

Font X, Epler-Wood M (2007) Sustainable tourism certification marketing and its contribution to SME market access. In: Black R, Crabtree A (eds) Quality assurance and certification in ecotourism. CABI, Wallingford, pp 147–163

Frey N, George R (2010) Responsible tourism management: the missing link between business owners' attitudes and behaviour in the Cape Town tourism industry. Tourism Manage 31:621–628

Gebben C, Gitsham M (2007) Food labeling: understanding consumer attitudes and behaviour. Ashridge Business School, Berkhamsted

Goodwin H (2005) Responsible tourism and the market. International Centre for Responsible Tourism, Occasional Paper 4 November
Gordon K (2001) The OECD guidelines for multinational enterprises and global instruments for corporate responsibility. Working Paper on International Investments 2001/5, OECD – Directorate for Financial, Fiscal and Enterprise Affairs
Harford T (2005) The undercover economist. Oxford University Press, Oxford
Holloway C (2002) The business of tourism. Heinemann Butterworth, London
Jarvis N, Weeden C, Simcock N (2010) The benefits and challenges of sustainable tourism certification: a case study of the Green Tourism Business Scheme in the West of England. J Hospit Tourism Manage 17(1):83–93
Kalish A (2002) Corporate futures. Social responsibility on the tourism industry. Tourism Concern, London
Levy SE, Duverger P (2010) Consumer perceptions of sustainability in the lodging industry: examination of sustainable tourism criteria. In: Paper presented at the International CHRIE Conference-Refereed Track, University of Massachusetts, Amherst 30 July 2010
Martínez P, Rodríguez del Bosque I (2013) CSR and customer loyalty: the roles of trust, customer identification with the company and satisfaction. Int J Hospit Manage 35:89–99
Miller G (2001) Corporate responsibility in the UK tourism Industry. Tour Manage 22:589–598
Mitchell VW, Vassos V (1997) Perceived risk and risk reduction in holiday purchases: a cross-cultural and gender analysis. J Euromarket 6(3):47–97
Nicolau JL (2008) Corporate social responsibility. Worth-creating activities. Ann Tourism Res 35 (4):990–1006
Nurn CW, Tan G (2010) Obtaining intangible and tangible benefits from corporate social responsibility. Int Rev Bus Res Pap 6(4):360–371
Pivato S, Misani N, Tencati A (2008) The impact of corporate social responsibility on consumer trust: the case of organic food. Bus Ethics: A Eur Rev 17(1):3–12
Prud'homme B, Raymond L (2013) Sustainable development practices in the hospitality industry: an empirical study of their impact on customer satisfaction and intentions. Int J Hospit Manage 34:116–126
Rome A (2007) Financing tourism certification programs. CESD - Center on Ecotourism and Sustainable Development, Handbook 4
Russillo A, Honey M, Rome A (2007) Practical steps for funding certification of tourism business. CESD - Center on Ecotourism and Sustainable Development, Handbook 2
Schoorman FD, Mayer RC, Davis JH (2007) An integrative model of organizational trust: past, present, and future. Acad Manage Rev 32(2):344–354
Swaen V, Chumpitaz RC (2008) Impact of corporate social responsibility on consumer trust. Recherche et Applications en Marketing (English Edition) 23(4):7–34
Tamajón LG, Font i Aulet X (2013) Corporate social responsibility in tourism small and medium enterprises evidence from Europe and Latin America. Tourism Manage Perspect 7:38–46
Vitell S, Paolillo J, Thomas JL (2003) The perceived role of ethics and social responsibility: a study of marketing professionals. Bus Ethics Q 13:63–86
Weber J (2007) Fair trade coffee enthusiasts should confront reality. Cato J 27(1):109–117
White AL (2006) Business brief: intangibles and CSR. BSR - Business for Social Responsibility
World Bank (2003) Strengthening implementation of Corporate Social Responsibility on global supply chains. http://siteresources.worldbank.org/.../CSR/Strengthening_Implementatio.pdf. Accessed 12 Jan 2010
World Business Council for Sustainable Development (1999) Meeting changing expectations: Corporate social responsibility. http://www.wbcsd.org/templates/TemplateWBCSD5/layout.asp?type¼p&MenuId¼MTE0OQ. Accessed 12 Jan 2010

Chapter 4
Introduction to Reporting Systems

Abstract This chapter introduces reporting systems. Firstly, it explains what reporting systems are and underlines the benefits they brought for companies and consumers. Secondly, it describes their main characteristics, paying particular attention to those programs that are specifically addressed to assess the responsibility of tourism companies. While quality systems have existed for decades, environmental and social programs have been developed starting from the end of the 1980s/1990s depending on the geographical region. Type of auditing (self-assessment, second-party or third-party audit) and type of standards (process-based, performance-based or product-based standards) are the main elements that distinguish reporting systems. In addition, for programs addressing the responsibility of tourism companies, another important aspect to take into account is whether the system focuses on the evaluation of the internal business process and the whole supply chain or of the product only, checking the standards a responsible tourism product should respect ("certification approach" vs. "responsible tourism approach"). Many of these systems are scarcely known by companies and consumers and this damages their credibility and application.

4.1 What Is a Reporting System?

A reporting system can be defined as a way of measuring, ensuring and communicating that a product, an activity, a process meets specific standards (Kalish 2002; Goodwin 2005; Bien 2008; Bobbin 2012a, b).

It is a marketing-based tool that links players in the value chain by guaranteeing or certifying to users (tourists and tourism intermediaries) that the suppliers (tourism providers) adhere to a specific set of standards and indicators (Russillo et al. 2007).

M. Manente et al., *Responsible Tourism and CSR*, CSR, Sustainability, Ethics & Governance, DOI 10.1007/978-3-319-06308-9_4,

There are many schemes regarding tourism that relate to sustainability (e.g. Green Globe, Blue Flag) and a number of other projects are in the pilot or development phase in different parts of the world. However, the attention to sustainability and triple bottom line is relatively new: while quality systems (e.g. "star rating" in hotels) have existed for decades, environmental programs have been developed starting from the end of the 1980s, and socio-cultural and economic aspects have been added since 1990s in the Americas and the 2000s in Europe (Bien 2008). The impetus for environmental reporting systems and certification began in particular with the 1992 Earth Summit, held under the United Nations sponsorship in Rio de Janeiro, and then with the Agenda 21. Since 2000, there has been a growing recognition that most environmental certification programs have not seriously considered socio-cultural issues, while social and labour systems have not considered environmental factors (Bien 2008).

In addition, in the tourism industry not all sub-sectors have shown the same awareness and interest in adopting these systems. Until the mid-2000, most initiatives (more than 60 %) were concentrated in the accommodation sector, while less than 20 % addressed destinations, 7 % tour operators and 5 % other tourism activities (sport and leisure, transport, etc.) (Dodds and Joppe 2005).

Programs and practices vary region to region around the world. In order to avoid the dangers of uncontrolled proliferation, some formal initiatives were established (see the Global Sustainable Tourism Council: http://www.gstcouncil.org/) in order to identify, recognize and give publicity to reporting systems with minimum standards to ensure sustainability.

These systems, if well implemented by companies, are considered to play an important role for the tourism industry. They are not an end in itself, but a tool for motivating businesses and other to improve their environmental, social and economic performance, while rewarding them for doing so. Therefore, reporting systems can bring benefits to (Bien 2008):

- Companies: reporting programs educate them and stimulate voluntary improvements; they tend to reduce operating costs, to support better operating business, to allow easier access to technical assistance and financing; they are thought to give tourism companies a competitive advantage and increased linkages into the distribution chain and to consumers through marketing;
- Consumers: reporting systems help them to environmentally and socially responsible choices, they help them to distinguish businesses that are truly responsible and to have better services, they increase public awareness and alert tourist to specific issues in a destination;
- Governments: reporting systems help them to protect their market niches, especially when the credibility of destinations is threatened by greenwashing; they support them to know to what extent the tourism development in their destinations is sustainable and how every stakeholder contribute to it;
- Environment and local communities: reporting systems indirectly protect both the environment and the social and economic structure of local communities,

since they require businesses to respect the local culture and to provide real social and economic benefits for it.

However, in spite of these opportunities, many of these systems have insufficient funds for effective, broad-scale consumer marketing that can be notoriously expensive (Bien 2008). Therefore many reporting programs are not easily recognisable.

In addition, as Goodwin states (2005), two further issues have to be taken into account. From the consumers' point of view, reporting systems are opaque: for example, although potential clients know that the company has awarded a certification label, they find it difficult to understand what kind of benefits this award can bring to them and, above all, what are the elements that distinguish companies adopting and non-adopting a reporting program.

The second issue is the proliferation of reporting systems related to sustainability: while a large number of local schemes is desirable, because it is a product of the cultural and ecological diversity of the world and of different issues and priorities (Goodwin 2005), this implies problems in terms of programs' recognition and credibility on the part of companies and consumers.

4.2 Main Characteristics of Reporting Systems

There are many types of reporting systems, as well as related awards and eco-labels. However, although different, they share some similar components. These are (Bien 2008):

- Voluntary enrolment by businesses;
- Well-defined standards and criteria;
- Assessment and auditing;
- Recognition and awarding the use of a logo;
- Periodic follow-up audits to renew the assessment and the certification;
- Continual improvement;
- Transparency;
- Participatory mechanisms to define standards.

As for the fairness and objectivity of a reporting program, they are ensured by the following elements:

- Non-discrimination: systems should be available to all applicants who meet the standards and whose activities are covered;
- Systems should be free of undue financial consideration and independent of business size and membership in groups and organisations;
- Standards should consist of clearly defined criteria, specifically related to what is being assessed;
- The auditing organisation should be able to make impartial evaluations and certification decisions, based on clearly defined criteria;

- The reporting system organisation should allow the participation of all parties significantly concerned with the activity being certified, with respect to the content and functions of the certification system;
- The person who evaluates a business should be separate from the entity that decides whether or not to award the system and both should be independent of whoever provides consulting service to help the business qualify;
- The reporting body should be free from commercial or financial pressure that might influence decisions;
- The reporting should award a logo and include a sunset clause that requires re-application after a given time period, to protect against greenwashing;
- There should be established procedures for revising and changing criteria and standards and so on.

About the main distinctions among reporting programs, they basically concern:

- The auditing procedure (first- vs. second- and third-party audit);
- The focus of the reporting program (product- vs. process- based systems or process- vs. performance-based systems) (Bien 2008).

As for the auditing procedure, while first-party audit is basically a self-evaluation, second-party audit implies that a consumer or an industry body assures that the product meets the minimum standards. Third-party audit means instead that a neutral, independent third party evaluates the compliance of the product or of the company with clearly defined standards.

As for the type of standards, while product standards reflect the appropriate characteristics a product is expected to have, a controversial topic is whether processes or performances should be assessed and eventually certified. Process standards (e.g. ISO 9000 or 14001) refer to the appropriate characteristics of the process put in place, from the creation of the product to final distribution. They assess and certify businesses that have established and documented systems for assuring the improvement of quality or environmental and social performance. However, they do not determine any specific performance results other than those required by law. Consequently, businesses receive a certification for their efforts, not for their actual performance.

On the contrary, performance standards assess and certify whether or not a business or activity complies with objective external criteria. This allows a direct comparison between two businesses to show which one has better results (e.g. how many litres of water per guest per night the two hotels consume?). They measure achievements and results according to tangible criteria, not intent. They are more transparent and can include environmental and socio-economic criteria within and without business. They can offer different levels of logos reflecting different levels of performance.

Most of the available reporting schemes seem to be process-based and to focus on the awareness of the company about its impacts, but not on the results it achieves (Bobbin 2012a). However, process-based systems ensure that a company documents its performance, and more importantly, that it has internal controls in place in order to continue to maintain good performance. Whether a performance-based system can certify on the basis of one or a few measurements of key indicators, a process system can ensure that the indicators are constantly monitored and that there is a person and a budget available to do so.

Since both standards have their own pros and cons, there is the growing consensus in tourism that systems should incorporate elements of both process and performance standards (Bien 2008).

Currently, large tourism operators, like hotel chains, airlines and big tour operators, generally opt for international process standards such as ISO 9001, which provides a quality assurance accreditation, or ISO 14001, which provides specific environmental accreditation. One of the advantages of these schemes is that "they brought with them improved management practices and processes, although they were see as bureaucratic by smaller enterprises" (Goodwin 2005).

Starting from 2010, the ISO 26001 is at work in Europe. The International Organization for Standardization has decided to launch an International Standard providing guidelines for Social Responsibility (SR). The guidance standard has voluntary to use and, unlike other standards, it does do not include requirements and thus it is not a certification standard.

In addition to aspects discussed above, which are common to all reporting programs, systems assessing in particular the business responsibility can be differentiated in relation to the approach followed. According to Goodwin (2005), it is possible to identify and distinguish the "certification approach" and the "responsible tourism approach" (Table 4.1). They are not mutually exclusive in principle, but they imply a different perspective in assessing tourism company responsibility. In more detail:

- Schemes based on a certification approach apply a process- and/or performance-oriented perspective and are adopted in particular by businesses engaged in assessing their internal processes and supply chains and improving their management. However, this approach rarely provides tourists with an enhanced experience, since it is often difficult for consumers to check the effects of certification on product quality and to monitor the company improvement year on year.
- Schemes based on a responsible tourism orientation adopt a market-driven perspective. They focus on product standards, reflecting the appropriate characteristics a product of responsible tourism is expected to have. This approach responds to tourists who demand a more real encounter with the environment and the community, based on values of respect for other people and their places. These tourists are informed consumers who continuously review the products and recommend them to others when they are satisfied. This approach both responds to changing consumer demands and enables people to experience the difference, creating demand for new products.

While the first approach does not encourage products' differentiation and does not bring significant marketing advantages, the second perspective contributes to distinguish responsible tourism products from others and to produce benefits for companies, who are able to better answer to and attract responsible tourists.

Table 4.1 Comparison between the certification approach and the responsible tourism approach (Goodwin 2005)

	Responsible tourism approach	Certification approach
Driver	Market led	Supply-side governance and CSR
Users	Consumers	Investors and Businesses
Auditing	Consumer auditing	Third-party auditing
Orientation	Output orientated	Process orientated
Transparency	Highly transparent and content rich	Opaque—devoid of content
Consumer engagement	The enterprise makes some specific statements about the experience—part of the contract	Low—the consumer know only that a certification has been awarded
Focus	Economic, social and environmental	Primarily environmental
Funding	No subsidy—self-financing	Expensive—subsidised
Legal	Contractual obligation	Very limited liability
Improvement	Consumers expect to see the improvement	Certified commitment to continuous improvement
Marketing value	Strong marketing value—high levels of product differentiation focused on experiences of particular places and communities	Very weak marketing value—no differentiation between products and experiences
Competition	Significant—there is a ratchet effect as a result of competition and increasing consumer pressure	Limited—tends to produce a level playing field

However, responsible tourism and certification approaches are not incompatible or mutually exclusive. The certification approach is more useful as a means of ensuring compliance and improvement in the supply chain, while the responsible tourism perspective is more appropriate for engaging and reassuring responsible tourists (Goodwin 2005). Therefore, a combination of both approaches seems to be a valid solution.

References

Bien A (2008) A simple user's guide to certification for sustainable tourism and ecotourism. CESD - Center on Ecotourism and Sustainable Development, Handbook, n. 1

Bobbin J (2012a) Researching responsible tourism reporting. International Centre for Responsible Tourism, Occasional Paper OP. 24

Bobbin J (2012b) Introduction to responsible tourism reporting. The development of a transparent verifiable reporting system suitable for small to medium sized enterprises to monitor their responsible tourism good practices. Responsible Tourism Reporting. http://www.rtreporting.org/about/news-and-publications.php. Accessed 8 Aug 2013

Dodds R, Joppe M (2005) CSR in the tourism industry? The status of and potential for certification, codes of conduct and guidelines. CSR Practice Foreign Investment Advisory Service – Investment Climate Department, Washington DC

Goodwin H (2005) Responsible tourism and the market. International Centre for Responsible Tourism, Occasional Paper 4 November

Kalish A (2002) Corporate futures. Social responsibility on the tourism industry. Tourism Concern, London

Russillo A, Honey M, Rome A (2007) Practical steps for funding certification of tourism business. CESD - Center on Ecotourism and Sustainable Development, Handbook 2

Chapter 5
The Analysis of European Reporting Systems for Responsible Tourism and CSR

Abstract This chapter proposes a detailed overview of reporting systems for assessing tourism business responsibility, in terms of compliance with responsible tourism values or CSR principles. The analysis focuses, in particular, on programs that have been mainly developed and applied in Europe and addressed to tour operators. The analysis includes eight reporting systems: the Responsible Tourism Standards of the Italian AITR; the system of the French ATES; the Enterprise Indicator for Responsible Tourism of the Spanish QUIDAMTUR; the CSR Reporting Standards of the German KATE and TourCert; the ATR system of the French ATT; the Tour Operators' Sector Supplement of the international initiatives TOI and GRI; the system of the English Responsibletravel.com and the Travelife Sustainability System of the European initiative Travelife. The main characteristics of every system have been described (objectives, approaches, standards and indicators, auditing procedure) and a final comparison has been developed in order to identify main commonalities and differences. In addition, the main issues related to these programs have been identified and discussed: the difficulties met by small- and medium-sized enterprises in implementing them; the different attention paid to the triple bottom line dimensions; the inactivity of some systems versus the great consensus reached by other programs.

5.1 The Methodology

In this chapter a study of the main reporting systems developed in Europe to assess responsible tourism or CSR has been carried out. The aim is to analyse and compare the different programs in order to identify common points and differences and to discuss the main issues associated to their implementation by tourism companies.

Four specific criteria have been used to select the programs to be included in the analysis.

M. Manente et al., *Responsible Tourism and CSR*, CSR, Sustainability, Ethics & Governance, DOI 10.1007/978-3-319-06308-9_5,

1. Programs that monitor and assess the business responsibility along the overall triple-bottom line, i.e. according to the social, environmental and economic dimension. Therefore, eco-certifications or other systems focussed on one dimension only have been excluded. As explained in the previous chapters, this decision has been taken because both responsible tourism and CSR require companies to take responsibility for the economic, social and environmental effects produced by their actions;
2. Programs that have been developed and promoted by European organisations or that are mainly applied within the European tourism industry. Given the complexity of mapping the overall international context, the decision was taken to focus on Europe.
3. Programs that address in particular tour operators. Given their central role in the tourism industry and in the whole supply chain, tour operators interact with all businesses involved in the organisation and assembling of the travel package. Consequently, responsible tour operators can also stimulate other companies to behave in a responsible way. In addition, not only the assessment of the tour operator's responsibility gives consumers information about the tour operator itself, but also about other companies along the supply chain that offer the services included in the travel package.
4. Programs for which detailed and comprehensive information are available to the public.

According to these four criteria, eight reporting systems have been included in the analysis, as explained in Table 5.1.

While some of these schemes have already been implemented by tour operators (e.g. TourCert and Travelife), others are still in a development phase or have not found a practical application in the market yet. The latter is the case for the schemes developed by AITR, QUIDAMTUR and TOI-GRI. In addition, as it will be better explained in the next paragraph, the application of some programs are limited to a specific country (e.g. ATES and ATR) or is available only for the associate travel agencies and then to assess their partners and suppliers (Responsibletravel.com). Although these limitations, these reporting systems have been included in the analysis since they can offer interesting cases for discussion on the approaches and the indicators used.

5.2 AITR: Responsible Tourism Standards

5.2.1 Organisation and Objectives

The Italian AITR (Associazione Italiana Turismo Responsabile, in English Association for Responsible Tourism) is a non-profit network, with about 200 members, 50 of which are tourism SMEs (small travel agencies, tour operators and accommodation establishments).

The Association has developed the Responsible Tourism Standards, with the aim at establishing the principles and values useful to certify travels made by their members with the label of "responsible tourism".

Table 5.1 Reporting systems included in the analysis

Name of the reporting system	Organisation/Association	Country
Responsible Tourism Standards	AITR (Associazione Italiana Turismo Responsabile—Italian Association for Responsible Tourism)	Italy
ATES system	ATES (Association pour le tourisme equitable et solidaire—French Association for Fair and Solidarity Tourism)	France
Enterprise Indicator for Responsible Tourism	QUIDAMTUR	Spain
CSR Tourism certified	KATE (Kontaktstelle für Umwelt and Entwiklung Centre for Environment and Development) Tourcert	Germany
ATR program	ATT (Association des Opérateurs Thématiques—Association of Thematic Tour Operators)	France
Tour Operators' Sector Supplement	TOI (Tour Operator Initiative) GRI (Global Reporting Initiative)	European Level
Responsibletravel.com System	Responsibletravel.com (online travel agency)	UK
Travelife Sustainability System	Travelife	European Level

In particular, the main goals are:

- Distinguish responsible tourism from other forms of tourism;
- Protect the consumer, ensuring him/her that the travel he/she buys respects the characteristics and principles of responsible tourism;
- Protect the travel organisers who are members of AITR from the improperly use that other operators can or could do of the term "responsible tourism", for marketing reasons only;
- Provide to AITR members a set of minimal operational standards and indications for the building of a responsible travel package (Fig. 5.1).

5.2.2 The Reporting System

The Responsible Tourism Standards applies only to travel packages created by AITR members; these are classified in two different groups—one for travels in Italy and the other for travels abroad—in order to take into account the peculiarities of different destinations (in terms of travel organisation, local partners, population, etc.).

The Standard involves small tour operators who are members of AITR, their partners and suppliers in the destination. In particular, it is granted by an internal auditor commission, to operators who:

- Are members of AITR;
- Have not been involved directly or indirectly in tourism activities (also intermediary activities) that are against the principles of responsible tourism or that

Fig. 5.1 AITR's Logo

imply pornographic purposes, exploitation of child labour, ravages of the environmental heritage and landscape.

The Standard is applied to those members who submit for certification a number of travel packages corresponding to at least 50 % of total turnover deriving from direct selling of travel packages themselves. If the travel package is organised respecting the criteria mentioned above, it will receive the AITR Standard and the company will be enrolled into the AITR Registry with the list of the travel products certified.

The methodology for the definition of the travel standard developed by AITR is based on five principles, divided into 13 criteria and 23 indicators, which verify if the product is conceived and organised in full compliance with the values of responsible tourism. Eight document sheets have to be provided by applicants to accomplish these criteria.

Table 5.2 shows all five main principles that all stakeholders involved have to respect and that reflect the values of responsible tourism. The principles underline the central role of the local community, the interaction between tourists and local people, tourists' awareness and education, the reduction of negative impacts, the fairness in the distribution of benefits.

The information useful to assess whether these principles are respected by each member are collected through a set of technical documents prepared by the applicant. These documents are listed in Table 5.3.

As mentioned before, for every principle a number of criteria have been developed, explaining the different aspects of each principle. These criteria have been translated by AITR into a set of indicators, which help auditors to test that the choices and actions made by the travel operators in organising the travel package are coherent with the principles of responsible tourism. The testing is made according to the information collected through the set of documents described above.

Table 5.4 shows, for every principle, the criteria and the indicators identified and the sources used to build them.

Table 5.2 Five principles of AITR responsible tourism standards

Principle 1: Enhancing the centrality of the host community

- Responsible tourism recognises the centrality of the host community and its right to be the driver of the sustainable and socially responsible development of its territory
- It acts supporting the positive interaction between the tourism industry, the local community and travellers
- It respects and encourages the right of local communities to decide the tourism development in their territory and establishes continuous fair-trade co-operation. Local communities can express an opinion about the tourism movements, deciding whether to welcome visitors or not and, if yes, choose appropriate welcoming proposal

Principle 2: Supporting the meeting with cultural backgrounds and the knowledge of the historical and environmental heritage of the destinations visited, respecting the differences

- Responsible tourism acts to draw growing attention on the interaction between tourists, tourism industry and host community, respecting cultural difference
- It proposes travel solutions which allow travellers to interact with places and people living in the destinations visited and to deepen the knowledge of the historical, cultural and environmental heritage
- It favours relationships between travellers, people and local communities, by creating opportunities for meeting and exchange
- It asks the traveller to adapt to local customs, in the light of mutual respect and equal dignity between cultures and customs

Principle 3: Developing travellers' awareness

- Responsible tourism aims that travellers become conscious of their role as consumers of a responsible tourism product, which affect the quality of tourism supply and the lives of millions of people residing in the destinations
- It asks the traveller to assess his/her real travel motivations and the travel organisers to prepare tourists before leaving, in order to make them aware of the characteristics and peculiarities of the travel he/she is going to do
- The traveller is asked to think about the environmental, social and economic impacts of his/her travel also during and after the travel itself

Principle 4: Minimising environmental and social impacts

- Responsible tourism aims at minimising socio-cultural and environmental impacts produced by tourist flows and activities
- It acts supporting choices that imply a minimum environmental impact, asking both local communities and travellers to become aware of different travel impacts in order to guide their behaviour
- It proposes a critical remark on means of transport at high environmental impact, like air flights, promoting the travellers' consciousness about this issue and carrying out compensation policies

Principle 5: Developing solidarity economy/equity/legality

- Responsible tourism is tourism developed according to the principles of economic and social justice, respecting the environment and culture
- It makes sure that the price paid is equitable for all (tourists, operators and local communities) and that the income produced also benefit other sectors and have positive social impacts at destination
- It favours the use of local small-scale or family-managed services (accommodation, restaurants, transports, etc.) and prefers those services that can be used by both visitors and residents
- It operates so that the actors involved in responsible tourism activities respect the regulations in force and there is an acknowledgment of the specificity of this kind of tourism by tourism authorities

Table 5.3 Technical documents required by AITR Responsible Tourism Standards

1. Travel technical sheet It is prepared by the applicant for the travellers and delivered before the travel. It includes: • Flight plan and information regarding travel insurance; • Presentation of the travel itinerary; • Travel program day-by-day; • Presentation of local partners and of projects visited during the proposed itinerary.
2: Travel internal technical report It is prepared by the applicant for internal use. It includes: • Report on travel conception and planning; • List of partners and providers (updated annually) ; • Costs and feed backs from partners (updated annually).
3: Standard sheet on "price composition" It states the economic impact of the responsible travel on local community, expressed by the price share (in %) going to the benefit of resident population.
4: Questionnaire on customer satisfaction It is a questionnaire prepared by the applicant and filled in by the traveller once come back home. Travellers can further contribute to the monitoring process of certified travels by sending critical reviews of their travel experiences made with AITR members. These reviews can be sent using a standard format prepared by AITR that travellers can download directly from the AITR website.
5: Administrative/legal documents It is a set of documents the travel organisers must provide by law. They include: • Tour operator licence and Chamber of Commerce enrolment; • Alternatively, statute and enrolment to the national APS register; • Insurance policies the travel organisers should have by law; • Travel contract.
6: Local partner sheet It is prepared by the applicant for internal use. It includes: • Partner personal data and description of his activities; • Documents stating the relationships with partner organisations. Where there are no formal contracts, if required, the travel organiser provides the auditor the mails exchanged with partners and providers.
7: Travel preparatory sheet. It is developed by the organiser and it self-certifies the information/training activities done for the traveller before–during–after the journey. The sheet includes at least: • An informative telephone call before booking; • Delivery of information materials (also by email); • A preparatory meeting before the journey with a representative of the travel organiser; • Monitoring during the travel; • Checks after the travel (see Sheet n. 4 and other initiatives).
8: "Minimising environmental impact" sheet It is prepared by the travel organiser, which self-certifies that the choices made for the travel organisation minimise the environmental impact of tourism on the destination. It includes: • The use of KM 0 products; • Destination's carrying capacity; • Energy savings; • Transports and emissions; • Water consumption; • Waste disposal.

Table 5.4 Criteria and indicators of AITR responsible tourism standards

Principle 1	
Criteria 1.1—The travel is organised with the participation of local communities	**Documents**
Indicator 1.1.1—The travel organiser plans and/or manages the whole travel or part of it in co-operation with local partners/providers or other people active at local level	1, 2, 6
Indicator 1.1.2—The travel organiser uses services provided by local tourism operators	2
Principle 2	
Criteria 2.1—Meetings with the local population are a fundamental part of the travel experience provided	
Indicator 2.1.1—The travel organiser uses local guides/cultural agents	1, 2, 4
Indicator 2.1.2—The travel program includes some meetings with local testimonials and associations	1, 2, 4
Criteria 2.2—The travel program pays attention to local diversities and cultural peculiarities	
Indicator 2.2.1—The travel program proposes the tourist to experience different cultural backgrounds	1, 2, 4
Criteria 2.3—The travel program pays attention to natural and environmental characteristics of territories	
Indicator 2.3.1—The travel program proposes tourists the direct knowledge of elements characterising the biodiversity of territories and, where possible, of initiatives of nature preservation	1, 2, 4
Criteria 2.4—The travel is flexible and organised in a way that allow tourists to deepen the knowledge of the destination	
Indicator 2.4.1—The travel organiser commits himself to propose programs that balance the travel length with the time of transfers	1
Indicator 2.4.2—Monitoring the travel program during the travel	4, 7
Principle 3	
Criteria 3.1—The travel organiser provides appropriate information to tourists and make them aware of the socio-economic and cultural aspects of the destination	
Indicator 3.1.1—The travel organiser prepares and delivers a "Travel technical Report" to the traveller (FD1). There should be an acknowledgment stating that the Report has been presented and delivered before the travel	1, 4
Indicator 3.1.2—The travel organiser assures the presence of a cultural agent, with proved expertise and knowledge of the destination and also of the definition of responsible tourism according to AITR principles	1, 2, 4, 6
Criteria 3.2—The travel organiser informs tourists about the values of responsible tourism	
Indicator 3.2.1—The travel organiser provides a preliminary preparation of tourists to responsible travels	7
Indicator 3.2.2—The travel organiser provides a preliminary meeting before the travel in order to make tourists awareness about the values of responsible tourism	7
Principle 4	
Criteria 4.1—The activities included in the travel program are organised respecting the local eco system and the habitats where they take place	

(continued)

Table 5.4 (continued)

Indicator 4.1.1—The travel does not take place in fragile environmental areas, except in case of special agreements with local authorities	8
Indicator 4.1.2—To minimise the environmental impacts, the travel organiser: identifies the critical areas included in the travel itinerary; collect the local regulation for preserved areas and do his best to respect it; adopts specific measures to minimise impacts; writes all these measures in the Report "Minimising Environmental Impact".	8
Criteria 4.2—The travel organiser chooses most suitable transport modes in order to minimise their use and the impact on the environment	
Indicator 4.2.1—Whether air transport has to be used, the travel organiser commits himself to compensate the environmental impact with the support to reforestation programs, to the development of renewable energies or to other initiatives proved to be effective in reducing carbon dioxide	8
Indicator 4.2.2—The travel itineraries have been adapted—when possible—to the timetable of local transports having limited environmental impact	8
Criteria 4.3—The travel organiser adopts approaches and techniques that minimise the environmental impact of tourist activities, taking into account the specific features of the travel	
Indicator 4.3.1—The travel organiser , according to conditions and rules established in travel destinations, develop suitable actions in order to limit waste production, such as avoiding the use of plastic tableware and plastic bottles (except for specific health conditions); using recycled paper for catalogues and brochures; using recyclable vessels for foods and drinks	8
Indicator 4.3.2—The travel organiser informs clients about the actions he has adopted to minimise environmental impacts	8
Principle 5	
Criteria 5.1—The travel is organised according to the principles of transparency and equity for all actors along the tourism value chain	
Indicator 5.1.1—The travel organiser shows that an adequate documentation has been prepared explaining the items included in the travel prices	3
Indicator 5.1.2—There is evidence that the Price Composition Sheet has been delivered to travellers before departure	3
Criteria 5.2—The travel has been organised so a so maximise positive impacts on the local economy and the community	
Indicator 5.2.1—Locally-owned and managed accommodation, restaurants and transports are used (where present), chosen according to the best quality/price relationship	1, 2, 3, 4, 6
Indicator 5.2.2—The travel organiser develops or promote and support (financially and/or with its organisation) the development of projects for the local community	-
Criteria 5.3—Travel organisation is provided by a company respecting the Italian law in the field of tourism	
Indicator 5.3.1—The travel organiser has the licence, the director and the insurance required by the law	5

The AITR Responsible Tourism Standards is based on a self-assessment procedure, carried out by the travel organiser, and a subsequent ratification by AITR through an internal group of auditors.

The system follows both a market-driven perspective and a responsible tourism approach, since the focus of the assessment is the travel package and not the tour operator activity. Through its Standards, AITR ensures consumers that the travel packages sold by its members respect the principles of responsible tourism; however it does not consider the internal tour operator responsibility, which is only indirectly assessed. For example, the Standards does not include any indicator about the human resource management or the responsibility along the business processes and the whole supply chain; about this last aspect there are no indicators that verify if partners and suppliers are responsible.

The AITR system seems to be more focussed on the social and cultural dimension of travel, than on the environmental and economic ones. The centrality of the local community and its participation to the organisation of travel, the meeting between tourists and the resident population, the respect of diversities are key elements of its philosophy.

Finally, the system is based on a qualitative approach, since that quantitative indicators are not included and the assessment procedure does not contemplate to express the responsibility level through a numerical value, as requested by other reporting systems. This characteristic can depend on the greater complexity linked to a quantitative procedure, in particular for small tour operators, as the enterprises who are members of AITR. As a consequence, a qualitative system can make easier the adoption of the Standard by AITR tour organisers.

As previously mentioned, after its development and the discussion among AITR members, the Responsible Tourism Standards is not yet implemented by any tour operators of the association, who are even opting for other systems. Therefore it remains a theoretical experience and a reference point for AITR members but not an applicable reporting system.

5.3 ATES: Charte du Tourism Equitable et Solidaire

5.3.1 Organisation and Objectives

The French ATES (Association pour le Tourisme Equitable et Solidaire; in English, Association for Fair and Solidarity Tourism) was created in 2006 in France by the Union Nationale de Tourism et de pleine air (UNAT), the Plate-Forme pour le Commerce Equitable (PFCE), the federation Loisirs Vacances tourisme (LVT) and travel professional associations. ATES groups nearly 22 tour operators selected according to a grid of ethic criteria and that commit themselves to develop tourist activities that helps the local development of destinations, through a tight co-operation with local communities and their representatives (Fig. 5.2).

The association was formed according to a common Chart and represents the first national network of tour operators specialised in responsible tourism in France.

Fig. 5.2 ATES's Logo

Like AITR, whatever the destination chosen, tourists who choose a responsible package sold by ATES members, travel in small groups, away from big tourism infrastructures. They experience special contact with local populations, respecting their culture and environment, they stay in the habitants' homes or in accommodation close to the villages and they contribute to the local economy. A part of the price paid by tourists is used to finance development projects at local level, which are decided and managed by the local community directly.

The "Chart for a fair and solidarity tourism" summarises the main principles that the tour operators who are members of ATES have to respect, starting from two fundamental concepts on which responsible tourism is based:

- Tourism is an important driver of development for local communities and destinations;
- The project developed—i.e. the tourism activity carried out by the tour operator and the development program linked to it—is planned and managed in tight co-operation with local communities and their representatives.

5.3.2 The Reporting System

The system developed by ATES is applied to the structure and the organisation of the tour operator who are member of the association and not to the products/travels, as required the AITR program.

It is organised along three main axes that are defined in the Chart mentioned above, 10 criteria and 29 indicators (Table 5.5).

As AITR, ATES asks tour operators who apply the system to present a list of documents proving the respect of the principles included in the charter (Table 5.6).

The three axes are developed into the following 10 selection criteria and 29 indicators, described in Table 5.7. The grid is used to select the tour operators who can be invited or who can apply to become active members of ATES. It certifies the context in which the tour operator works and the content of the responsible tourist activity, i.e. the quality of the organisation process and of the partnerships put in place to arrange the travel to the destination.

Table 5.5 3 ATES. Axes of the "Chart for a fair and solidarity tourism"

Axe 1: The relationship with local partners/providers and the travel organisation

- The tour operator organises the stay at the destination in co-operation with the local communities, who are the driver of the welcoming process. It also favours the meeting and the exchange between the local communities and tourists as the core of the travel
- The tour operators gives priority local partners organised around a development project that goes to the benefit of the local community
- The tour operator organises its travel packages in order to maximise the positive impact on the destination, respecting its environmental, economic and social equilibrium

Axe 2: The involvement in favour of the local development

- The tour operator organises its activities and fix the price of its travel packages so as to allocate resources to development funds
- These funds are given to the populations at the destination

Axe 3. The travellers' awareness and the transparency in communication

- The tour operator informs travellers and make them aware about the principles of voyage solidaire and the characteristics of each destination sold (economic, environmental, social and cultural situation, manners)
- The tour operator informs the traveller about the price sharing of its travel packages
- It communicates to its clients about the actions carried out for the local development and the environment preservation
- The tour operator promotes fair and solidarity tourism and the respect of its principles. It also promotes the ATES network and its activities

Table 5.6 Technical documents required by ATES system

1. Brochure with the travel description
2. Description of other communication tools used (e.g. information on the tour operator's website and its contents)
3. Customer satisfaction questionnaire, filled in by tourists
4. Report on travels' conception, planning and organisation
5. Partnership agreements
6. Partners' presentation
7. Report on the activities carried out by the tour operator and, if required, by partners
8. Financial report of the tour operator and, in required, of its partners
9. Study on the impact of travels on local communities and destinations, also carried out by external experts
10. Travel plans, with the indication of the size of the group
11. Copy of the chart
12. Graphic tool showing the composition of travel price
13. Documents delivered to travellers
14. Destination sheets
15. Technical sheet by country or region
16. Other documents considered useful for the assessment

Table 5.7 Criteria and indicators of ATES system

Criteria 1—The local population is at the earth of the welcoming process and the travel organiser arrange the tourists' stay with its co-operation	Documents
Indicator 1.1—The travellers are welcomed by the local population and are in contact daily with them	1, 3, 16
Indicator 1.2—The local partners are informed about tourists' feedbacks	16
Indicator 1.3—The tourists' stay has arranged in co-operation with the local population	4, 16
Indicator 1.4—The tourists' stay is part of a partnership project and partners are clearly identified	5, 16
Criteria 2—The travel operator chooses partners that are involved in development project for the local community	
Indicator 2.1—The local partner works for the interest of local community	3, 6, 16
Indicator 2.2—The tourists' stay allow local development projects to be financed	3, 8, 16
Indicator 2.3—The development project involves a network of local operators and associations. These actors are clearly identified	7, 9, 16
Criteria 3—The travel operator organises its travel programs in order to maximise positive impacts on the local economy	
Indicator 3.1—During the stay, the travel operator encourages the use of local resources by the travellers and highlight the abilities of local population and the characteristics of local activities	3, 16
Indicator 3.2—The local population is employed for the organisation of the tourists' stay	3, 7, 16
Indicator 3.3—The travel operator controls that work conditions respect the labour international rules so and that wages given to each partner's worker is fair	7, 16
Indicator 3.4—The travel operator helps to organise training activities for the partners, if needed	7, 16
Criteria 4—The travel operator organises its travel programs in order to respect the local economic, social and environmental balance	
Indicator 4.1—The travel operator provides information on the population's quality of life and on the environmental situation of the destination	16
Indicator 4.2—The travel operator controls that the tourists activities it has developed at destination do not create an irreversible economic dependency from tourism	16
Indicator 4.3—The travel operator adopts some actions with its local partners and clients in order to minimise their own impacts	2, 3, 16
Indicator 4.4—The travel operator manages the number of tourists in each destination, according to the area and the seasonality, in co-operation with local partners	10
Criteria 5—The travel operator promotes the meeting between travellers and local communities during the travel	
Indicator 5.1—The travel operator favours the travellers' stay close to the population or in the residents' homes	1, 3, 16
Indicator 5.2—The travel operator arranges specific time for meetings between travellers and local population	1, 2

(continued)

Table 5.7 (continued)

Criteria 6—The travel operator promotes responsible tourism and its principles and engages himself in spreading ATES philosophy and activities	
Indicator 6.1—The travel operator declares publicly its engagement in favour of responsible tourism and recall its principles on its communications tools (brochures, website, etc.)	1, 2
Indicator 6.2—The travel operator commits himself to respect the ATES Chart and spread its principles among its travellers	3, 11, 16
Indicator 6.3—The operator arranges actions to make the public opinion aware of responsible tourism through conferences, interviews, exhibitions, etc.	10, 16
Criteria 7—The travel operator declares the composition of the travel price	
Indicator 7.1—The operator uses a graphical tool showing that shows the price composition by item (% share for international flights, internal expenditures of the operator, development fund and the remittances to local partners)	12
Indicator 7.2—The travel operator uses a graphical tool similar to the previous one, which does not take into account international flights	12
Indicator 7.3—The travel operator publishes the price composition and communicate it to travellers and public in general	2, 13
Criteria 8—The travel operator informs travellers on the development activities carried out at the destination and/or on the environment preservation actions developed; thanks to the tourist activities	
Indicator 8.1—The travel operator has documents certifying the development activities or preservation actions carried out by its partners	16
Indicator 8.2—The travel operator communicates the outcomes of the development and preservation actions during interviews, conferences, etc.	2, 7
Criteria 9—The travel operator informs travellers on the destinations visited (environmental, economic and social situation, behaviour rules)	
Indicator 9.1—The travel operator has information tools (brochures, documents, etc.) for each destination	14, 15
Indicator 9.2—The travel operator delivers to travellers a document such "the traveller guide", stating the commitments of travellers at the destination	3, 13
Criteria 10—The travel operator organises its activities and decides the travel price so as give part of it to a development fund reserved to local projects	
Indicator 10.1—Part of the development fund is devoted to the development of individual or general-interest projects (transferred from the operator from its margin or integrated in the remittance to local partners)	7, 12
Indicator 10.2—The travel operators or its partners arrange a specific procedure to select the projects that will be financed through the development fund	16

To the list of these 10 criteria, another parameter, that represents a pre-requisite, is added. This asks the tour operator to adopt an internal responsible management approach in terms of work environment, staff satisfaction, environment preservation, etc.

Operatively, the system, that is applied to both new entrants and current members, is based on a self-assessment procedure, then ratified by an internal commission, formed by ATES members and qualified external experts. The assessment concerns the travel programming on 50 % of destinations sold, with a minimum of three destinations (for tour operators programming more than three destinations), chosen among those having the highest number of travellers.

The assessment focuses on a project (the tourism activity developed by the TO and the related development program at local level) for each destination. Each member or candidate member has to self-assess this partnership filling in the grid and providing the set of required documents to ATES.

The system requires an evaluation that is based on a ternary scale: 0 = the behaviour of the operator does not correspond to the indicator; 1 = the behaviour of the operator partially corresponds to the indicator; 2 = the behaviour of the operator corresponds to the indicator, with an allowance permitted. For each criterion, a total score is obtained by calculating the arithmetic mean of the scores given to each indicator included in it. Therefore the maximum score achievable for the 10 criteria is 20. The output of the assessment procedure can provide three results: total integration, partial integration and stand by. In detail:

- A score higher or equal to 16/20 represents a total integration of the tour operator into the network (complete coherence with the Chart principles);
- A score between 16/20 and 13/20 represents a partial integration. The tour operator is asked to make progresses on some criteria, but its integration in ATES is not called into question;
- A score lower than 13/20 represents a warning: the tour operator is insufficient on a number of criteria. Its integration in ATES is called into question.

Consequently, to submit their documentation to the ATES internal commission, members and candidate members are required to justify a minimum score of 13/20. Current members are asked to fill in the evaluation grid every 3 years. The system is presently applied by nearly 25 associates.

Although most indicators are qualitative—as in the AITR system—, the use of a numerical assessment makes the procedure more quantitative, allowing the tour operators to be ranked according to their level of responsibility and to have a clearer idea about the activities and processes to be improved. In addition, it is possible to easily compare the performance of different members.

Being focussed on the structure and internal processes of the tour operator rather than on the destination or on a specific range of products, this methodology is only partially market driven. However, it is close to the responsible tourism approach, in that the main objective is to identify and recognise tour operators working in full compliance with the values of responsible tourism. The system pays also attention to the specific projects carried out by operators with local partners and to the related development program at local level, for each destination. In addition, it is wrong associating this system to a certification approach, since the responsibility along the overall supply chain is only partially assessed.

Looking at the dimensions of the triple bottom line, the ATES system focuses on the social aspect—even more than AITR—ignoring the environmental and economic dimension. ATES requires to reduce impacts and to respect the balances in the destination, but it does not specify with appropriate indicators these two areas of the corporate responsibility. However, this phenomenon depends on the association objectives and on the type of tour operators who are members of ATES; indeed the organisation and its member are engaging in promoting fair and solidarity travels, giving priority to the cooperation and well-being of the local community and to the meeting between tourists and local people; all elements that fall under the social dimension.

5.4 QUIDAMTUR: Enterprises Indicator for Responsible Tourism

5.4.1 Organisation and Objectives

QUIDAMTUR is a consulting company based in Malaga who proposes systems and procedures to make tourism activities as much sustainable as possible, in co-operation with public administrations, travel agents, accommodation establishments, other tourism services and travellers. It also organises training courses and actions to increase the awareness of tourism operators and consumers regarding CSR and responsible tourism.

In cooperation with Ismalar Rutas (a tour operator specialised in responsible tourism) and the University of Malaga-Department of Tourism, QUIDAMTUR has developed an integrated system for assessing the sustainability of the tourism supply along the overall tour operator supply chain.

From the perspective of tourism marketing, the system has to accomplish two basic goals:

- Its objectives has to be easily understandable by demand and testing agents; this implies offering an unambiguous tool to ensure that the products bought by the tourism enterprises involved in the evaluation process accomplish the principle of responsible tourism;
- It has to be universally applicable, so as to assess the sustainability of tourist services developed in both the North and the South of the world, favouring the competition between them; this means that the variables chosen are objective and allow researchers to rank tourist activities according to their level of responsibility.

The system addresses the single tourism enterprises involved in the production of the tourist service, so to companies who are suppliers of a tour operator. In particular, four main groups of activities have been identified:

- Accommodation establishments;
- Restaurant services;
- Transports;
- Other activities (tourist guides)

Fig. 5.3 QUIDAMTUR's Logo

The system is based on a procedure finally aimed at developing a specific indicator, the Enterprise Indicator for Responsible Tourism, which is able to assess the level of responsibility of each company, who is a provider of the tour operator. From this point of view, the QUIDAMTUR system supports the tour operator in assessing its responsibility along the to the supply chain (Fig. 5.3).

5.4.2 The Reporting System

The Enterprise Indicator for Responsible Tourism is the result of a three-step evaluation procedure:

- Self-assessment by the company interested in marketing their products with the label. For each group of tourism activity, a list of specific questions has been prepared;
- Evaluation by clients, through a satisfaction survey carried out after the consumption; the questionnaire can be filled in online or on paper during the;
- Evaluation by external experts (local incoming tour operators and travel agencies), through a deep investigation of the characteristics and structure of tourism supply and services sold.

According to QUIDAMTUR a three-step assessment procedure implies interesting opportunities and threats. It encourages a direct involvement of both tourists and companies in the assessment process and, as a consequence, a greater awareness and a higher comprehension of their role in responsible tourism.

The assessment on which the Enterprise Indicator for Responsible Tourism is based has to be carried out according to four aspects:

- Social aspect, linked to the relationship between the company and the local community;
- Environmental aspect, about the environmental consciousness of the enterprise, the actions it carries out to respect the environment (saving water, energy, etc.);
- Economics aspect, in terms of the impact produced by the company on the local and global economy;
- Cultural aspect, about the impact produced by the tourism activity on the culture of the local community.

For each aspect, four different types of variables have been indentified in order to detail the analysis:

- Internal variables, explaining the relationships between the tourism enterprise and the society/local community, produced by the actions characterising the daily activity of the enterprise (e.g. relationships with workers; internal management, etc.);
- External variables, describing the relationships with external stakeholders (shareholders and clients, distributors and providers);
- Variables related to actions in favour of society/local population, explaining the relationships between the company and the society/local community, through initiatives different from the actions of daily activity (e.g. investment in social and environmental actions; projects for social and cultural integration);
- Variables for assessment and control.

Each variable is in its turn divided into a number of sub-variables, which are parameters that can been observed and measured directly and that represent the basic observation unit (in total 98 sub-variables). Each sub-variable is converted into one question to be answered by each of the stakeholders who provide the information (clients, company, external auditors). Starting from a previous assessment of the weight attributed to each stakeholder, the value of the information provided from each of them will be balanced in order to get the final assessment. The variables and sub-variables presented are common to all tourist services analysed.

Auditors can answer to these questions with 0 %, 25 %, 50 %, 75 %, 100 %, depending on the level of accordance of the behaviour/performance of the company to what the variable requires. The final score derives from the mean of values obtained by the answers (sub-variables) given by managers, clients and external auditors to the questions related to each variable.

The variables and sub-variables included by the QUIDAMTUR system are described in Table 5.8. It is evident that not all variables need to be explained by sub-variables and that not all questions can be answered by all three groups of auditors (company, clients and experts). For example, clients do not have enough information to judge if the company fairly pays his employees and suppliers.

The system developed by QUIDAMTUR seems to be based on the certification approach, since its strong orientation to process and performance standards. However, the important role given to customers in the assessment procedure and the aim to offer transparency and guarantees to consumers highlight that the system also embeds a responsible approach.

Comparing this reporting program with the previous ones, it is evident that there are no indicators referring to the involvement of local population in the development of the tourist product and to the role played by the meeting between travellers and the local community, which are acknowledged as important aspect of responsible tourism.

Table 5.8 QUIDAMTUR. Variables and sub-variables included in the Enterprise Indicator for Responsible Tourism

Social aspect		
Internal Variables	Internal communication (information to workers)	
	Workers are treated respectfully and no discriminatory actions are put in place	
External Variables	The enterprise provides social information to:	– Clients – Shareholders
	The enterprise run its business respecting social principles towards:	– Providers – Distributors
	The enterprise adopts a respectful behaviour (respect of human rights but also accessibility for all and non discriminatory treatment) towards:	– Providers – Distributors – Clients – Shareholders
Variables for actions in favour of society/local community	The enterprises invests in social actions	
	The enterprise develops social promotion actions to local population	
	The enterprise condemns unfair situations	
	The enterprise promotes its social integration	
Variables for assessment and control	The enterprise describes the social impact of activities carried out	
	The enterprise assess the social actions developed	
	The enterprise introduces social actions in its planning	
	External assessment	
Environmental aspect		
Internal Variables	Recycling, reuse and reduction of materials:	– Fungible – Non fungible – Water – Energy
	Sustainable and responsible use of:	– Fauna & Flora – Water – Renewable energy. – Fungible materials – Transports
	The enterprise provides environmental information and advice to its workers	
External Variables	The enterprise provides environmental information to:	– Clients – Shareholders
	The enterprise run its business respecting the environment (relationships with distributors and providers) with:	– Providers – Distributors
	The enterprises invests in environmental actions	

(continued)

Table 5.8 (continued)

Environmental aspect		
Variables for actions in favour of society/local community	The enterprise develops environmental promotion actions to local population	
	The enterprise condemns unfair situations	
	The enterprise promotes its environmental integration	
Variables for assessment and control	Studies on carrying capacity	
	The enterprise describes the environmental impact of activities carried out	
	The enterprise assess the environmental actions developed	
	The enterprise introduces environmental in its planning	
	External assessment	
Economic aspect		
Internal Variables	Legal status of the enterprise	– Capital source – Legal entity
	Level of economic feasibility	– Turnover – Continuity
	Proper human resource management	– Training – Motivation – Remuneration – Selection
	Proper process management	– Organisation chart – Functions – Internal processes – Positions – Contingency plans
	Service quality	– Infrastructures – Equipment – Safety – Comfort – Cleanliness – Maintenance
	Use of local products	
External Variables	The enterprise adopts fair and honest behaviour with shareholder and clients	– Client's needs – Quality/price – Client care
	The enterprise adopts fair behaviour towards:	– Providers – Distributors
Variables for actions in favour of society/local community	The enterprise invests in environmental programs.	
	It participates to business networks for local development	

(continued)

Table 5.8 (continued)

Economic aspect		
Variables for assessment and control	The enterprise describes the economic impact of activities carried out	
	The enterprise assess the economic actions developed	
	The enterprise plans its economic actions in the short, medium and long term	
	External assessment	
Cultural aspect		
Internal Variables	The enterprise respects local customs	– Food – Events – Rites and legends – Language – Popular wisdom – History – Music
	The enterprise uses material elements that respect local culture	– Architecture – Crafts – Arts
	The enterprise provides cultural information and advice to its workers	
External Variables	The enterprise provides cultural information to:	– Clients – Shareholders
	The enterprise runs its business respecting local culture (Distributors, Providers) with:	– Providers – Distributors
Variables for actions in favour of society/local community	The enterprise invests in cultural actions	
	The enterprise carries out cultural promotion actions to local population	
	The enterprise condemns unfair cultural situations	
	The enterprise acts in favour of the cultural integration	
Variables for assessment and control	The enterprise describes the cultural impact of activities carried out	
	The enterprise assess the cultural actions developed	
	The enterprise introduces cultural actions in its planning	
	External assessment	

On the other hand, the QUIDAMTUR procedure assesses some aspects concerning quality service, safety, comfort, cleaning, quality/price, which are usually very important for tourists.

Finally, the use of a three-level assessment system assures more objective results and the assignment of numeric values to variables and sub-variables makes it easy to assess the level of business responsibility. However, the methodology appears complex and labour intensive, in particular for SMEs.

As the Responsible Tourism Standards of AITR, the Enterprise Indicator developed by QUIDAMTUR is in stand-by and it is not at the moment applied by any tourism company.

5.5 KATE-TourCert: CSR Reporting Standards in Tourism

5.5.1 *Organisation and Objectives*

KATE—the Centre for Ecology and Development—is an environmental and development organisation based in Stuttgart. It works on issues of sustainability in the North and the South of the world, especially in Latin America, with particular emphasis on environmental, quality and sustainability management, CSR, sustainable tourism, development education for fair globalisation and fair tourism.

In co-operation with Church Development Service—Special Desk Tourism Watch,[1] forum anders reisen,[2] UNI Europa[3] and six pilot enterprises, KATE has developed the CSR Reporting Standards for tourism SMEs, which is now managed by TourCert, a non-profit organisation for certification in tourism.

The main goals of the system are:

- To go beyond the mere raising of consciousness by tour operators regarding the concept of sustainability and responsibility in tourism, providing the companies with a set of tools that allow them to build an integrated responsible strategy and to assess and monitor their performances;

[1] Tourism Watch (TW) is a special desk of the church development service (EED). It works in the field of education and solidarity on Third World tourism and cooperates with ecumenical partners to promote sustainable, socially and environmentally friendly tourism development, following the motto “Fair tourism—A matter of heart and mind”.

[2] forum anders reisen is a business association of small- and medium-sized tour operators aimed at promoting sustainable tourism. The members satisfy a comprehensive set of criteria. Every 2 years an external institution will evaluate whether the criteria are being met. The non-profit umbrella organization was founded in 1998 by 12 tour operators and now has more than 140 members. The mission of the association and its members is to promote forms of tourism that are environmentally sustainable, economically feasible and ethically and socially just.

[3] UNI Europa is the European regional organization of Union Network International (UNI) with about 7 million members. UNI Europa is recognized as a European Trade Union association by the European Trade Union Confederation (ETUC). The basic aim of UNI Europa is to build a social and democratic Europe. To ensure the social and democratic dimension of the European integration process, UNI Europa works on policies and action related to the institutions of the European Union. One of the most important aspects in this context is to represent the member organisations in the European institutions that influence the social, economic and cultural conditions of the member organisations and their members.

Fig. 5.4 CSR Reporting Standards Logo

- To give transparency on the social, ecological and economic impacts of business operations;
- To ensure comparability and reliability;
- To increase credibility among customers.

The CSR Reporting Standards, that measures the degree to which the objectives of CSR and the sustainability policy of the tourism enterprise are achieved, which implies:

- Measuring the responsibility within the company;
- Measuring the responsibility along the supply value chain (product development);
- Developing improvement programs.

It has been developed following the main requirements that a reporting system should respect:

- Trustworthness;
- Relevancy;
- Clearness;
- Comparability;
- Reliability (Fig. 5.4).

5.5.2 The Reporting System

The starting point of the CSR Reporting Standards of KATE was the guidelines for sustainability reporting issued by the Global Reporting Initiative (GRI). These guidelines offer comprehensive sets of indicators for all areas of sustainability, especially in international companies. GRI has also developed a set of sector-specific indicators for tourism that are so far hardly used (see Sect. 5.7). In particular, these guidelines are too complex for small and medium-sized companies and the collection of data is too costly.

Therefore, the aim of KATE is to create a sector specific CSR reporting system for tourism businesses that offers incentives for improvement and that is easy to be apply especially in SMEs. In addition, the CSR reporting program fulfils the conditions of an environmental statement according to EMAS (Eco-Management and Audit Scheme) for the certification of environmental or sustainability management systems. The required elements, such as an environmental policy, a CSR manager, a review and description of all significant environmental aspects, a description of environmental objectives and a report to be annually updated, fulfil the core requirements of a management system that follows the so-called PDCA management cycle: Plan—Do—Check—Act.

The CSR Reporting Standards that is addressed in particular to tour operators is structured into sections, which in turn are divided into sub-sections that include appropriate indicators for measuring and assessing the CSR in all its aspects. In total, the system is based on 83 indicators (Table 5.9).

Not only KATE has developed the CSR Reporting Standards, but it has also proposed a specific guide that support companies, in particular SMEs, to implement the reporting procedure. This guide identifies the main steps to be followed in order to adopt CSR and prepare the CSR Report. The entire process, that, according to KATE, takes 6 months, implies eight steps.

- Decide and plan: the whole process will succeed or fail depending on the decision and support by the top management. Appointing a CSR manager conveys a clear message regarding the importance of sustainability in the company and provides clarity as to who is in charge of the process. Another first step is to develop a preliminary report structure and to think about content.
- Kick-off workshop—internal communication: all employees should participate in the kick-off workshop. In this workshop, the structure of the report and the check lists will be presented, the time frame will be set and the assignment of tasks will be agreed upon.
- Data review: by using check lists and sustainability checks, data of relevance to the company will be gathered by those in charge in order to describe the current situation regarding the responsibility versus customers (assessment of customer information and customer satisfaction), employees (assessment of staff satisfaction) and society (starting from the stakeholders map). In addition, economic, social and environmental data will be collected together with the product responsibility along the supply chain (through the sustainability checks). A portfolio analysis should be developed to assess and prioritise different fields of action according to two criteria: "potential for improvement" (X axe) and "relevance for sustainability"(Y axe). Not all of these tools need to be used when developing the first sustainability report.
- Data assessment: once collected, all data will be assessed. If possible, data will be visualised to facilitate presentation. All results are to be incorporated into the given report structure.
- Workshop—Developing strategies: in a second workshop with the employees, the team will be informed of the results of the data review. The policy will be

Table 5.9 KATE-TourCert. Sections and indicators included in the CSR reporting standards

1	**Company profile**
1.1	**General information**
1.1.1	Information about location, ownership, shareholdings in other companies, local branches, product portfolio
1.2	**Mission statement**
1.2.1	The enterprise has a mission statement that includes sustainability aspect and that is published on the enterprise's website
1.3	**Management**
1.3.1	The enterprise has implemented a CSR management system which is appropriate regarding its type and size
1.3.2	Top management assumes CSR responsibility and defines a working unit with a CSR officer
1.3.3	Legal Compliance: Knowledge of and compliance with all relevant laws and regulations (including health, safety, work environment and anti-corruption)
1.3.4	The enterprise has standardised contracts or a Supplier Code of Conduct which are signed by all suppliers and which regulate general conditions regarding human rights, biodiversity, climate protection and working conditions
1.3.5	The enterprise has identified relevant human rights issues and risks and has implemented procedures that allow the enterprise to conduct its due diligence
1.3.6	The enterprise has signed the Code of Conduct for the protection of children from sexual exploitation in travel and tourism
1.3.7	The enterprise has developed and evaluated a stakeholder map
1.3.8	A working suggestion and improvement management is in place. The improvement program is updated annually
1.3.9	With every (re-) certification the enterprise presents a sustainability report with updated data
1.3.10	The 10 core indicators must be published on the enterprise's website (enterprises with less than 250 employees may leave out up to 3 core indicators)
1.3.11	Dynamic improvement of performance: At least 4 core indicators must reach a value which does not fall below 25 % of the average value (updated annually)
1.4	**Economic data**
1.4.1	Composition of turnover (main fields)
1.4.2	Development of turnover during the past 3 years
1.4.3	Turnover per employee
1.4.4	Return on sales
1.4.5	Structure of expenditure
1.4.6	Percentage of expenditure which stays in the destination
1.4.7	Contribution to local added value (indicator composed of the following criteria: 2.2.3.3, 2.2.3.7, 2.2.3.9, 6.1.6.5, 6.1.6.6, 6.2.7.3, 6.2.7.4)
2	**Travel offers**
2.1	**Product portfolio**
2.1.1	Total number of booked flights
2.1.2	Transparent presentation of travel offers
2.1.3	Number of travel offers (own offers and offers by partner agencies)
2.1.4	Number of individual and group travellers
2.1.5	Average length of stay of individual and group tours

(continued)

Table 5.9 (continued)

2.2	**Travel offer development**
2.2.1	*The enterprise checks and evaluates its product portfolio against the following criteria. The evaluation results in a CSR index for travel offer development*
2.2.2	**Basic criteria of product development**
2.2.2.1	Information provided in the travel description and during the advisory process are researched by us or by our partners in the destination
2.2.2.2	Planning and development of our offers are based on standardised criteria/requirements which contain essential CSR aspects
2.2.2.3	Selection of new suppliers is based on standardised criteria/requirements which contain essential CSR aspects, such as labour standards
2.2.2.4	We actively communicate these criteria/requirements to our suppliers
2.2.2.5	We systematically collect data on the satisfaction of our customers and deduce improvement measures
2.2.2.6	The enterprise has a code of conduct for activities in indigenous and local communities which was developed in cooperation and consent of the affected community
2.2.3	**Criteria regarding quality and sustainability of product development**
2.2.3.1	Travel offers are revised on a regular basis in order to reduce negative social and ecological impacts, especially for local communities
2.2.3.2	In sensitive regions the size of the group is in accordance with the ecological and social circumstances
2.2.3.3	We co-operate with partners who pay fair wages, maintain humane working conditions, normal working hours as well as adherence to at least the minimal norms as defined by labour legislation
2.2.3.4	We co-operate with partners who are actively committed to the protection of children from sexual exploitation
2.2.3.5	We co-operate with partners who actively contribute to environmental protection
2.2.3.6	We co-operate with or promote protected areas and other areas of high/sensitive biodiversity
2.2.3.7	By our travel offers we support local structures (private accommodation, local agencies, guides, drivers, etc.)
2.2.3.8	The dialogue with our suppliers is honest, fair and is an active form of two-way communication
2.2.3.9	By our travel offers we support social and ecological projects within the destination
2.2.3.10	As far as possible we choose the most energy-efficient and low-emission means of transport for arrival and departure
2.2.3.11	We avoid feeder flights and offer our customers arrival and departure by train
2.2.3.12	We do not offer avoidable connecting flights within the destination
2.2.3.13	We choose the most energy-efficient and low-emission means of transport within the destination
2.2.3.14	We do not offer activities which have an extremely negative impact on the environment and on resources (off-road trips, tours by off-road vehicle, motorcycle or snowmobile, sightseeing flights by airplane or helicopter, heliskiing, etc.)
2.2.3.15	In sensitive (protected) areas our customers are guided by specially trained and authorised guides or rangers
2.2.3.16	The enterprise does not use tourist infrastructure that limits or deprives local communities of their livelihoods or of access to scarce resources, transportation, rights-of-way and housing

(continued)

Table 5.9 (continued)

2.2.3.17	Intercultural encounters, especially when happening in deprived areas or social hotspots are prepared accordingly
2.2.3.18	We co-operate with certified suppliers (environment, quality, social)
2.2.3.19	Natural habitats and the behaviour of wildlife is not being influenced in a negative way by our travel offers
2.2.3.20	Negative impacts on eco-systems are being minimised systematically and possible damage is being rectified
2.2.3.21	In protected areas fees are being paid to the reserve administration
2.3	**CO_2 emissions of travel offers**
2.3.1	Total CO_2 emissions of transportation and accommodation
2.3.2	Total CO_2 compensation
2.3.3	Total CO_2 emissions less compensated emissions
2.3.4	CO_2 emissions/guest/day less compensated emissions
2.3.5	Percentage of climate compensation
2.3.6	Compensated CO_2 emissions
2.3.7	Total flight kilometres
2.3.8	Average flight kilometres per bed-night
2.4	**Transport**
2.4.1	Total CO_2 emissions of transportation
2.4.2	CO_2/km per means of transport for arrival and departure as well as for transportation in the destination
2.4.3	Total CO_2/km for arrival and departure as well as for transportation in the destination
2.4.4	Flight kilometre per guest
2.5	**Bed-nights**
2.5.1	Number of bed-nights per accommodation category
2.5.2	kWh per accommodation category
2.5.3	CO_2 per accommodation category
2.5.4	Total number of bed-nights
2.5.5	Total kWh for bed-nights
2.5.6	Total CO_2 emissions (tons) for bed-nights
3	**Customers**
3.1	**Customer satisfaction**
3.1.1	Index of customer satisfaction (annual evaluation of customer feedback)
3.1.2	Response rate of customer feedback
3.1.3	Guide to tourist ratio (group size per guide)
3.2	**Customer information**
3.2.1	*The enterprise checks and evaluates its customer information against the following criteria. The evaluation results in a CSR index for customer information*
3.2.2	**Basic criteria of customer information**
3.2.2.1	The company has a manual/guidelines for the advisory and booking process, which the employees comply with
3.2.2.2	A customer feedback system has been installed and improvements are implemented systematically

(continued)

Table 5.9 (continued)

3.2.2.3	There is a contact person in the destination who can be reached by telephone at any time (emergency number)
3.2.2.4	CO_2 emission of every travel offer are accounted and publicly listed
3.2.2.5	The enterprise actively promotes CO_2 compensation by the customers
3.2.2.6	Customers receive information about the Code of Conduct regarding the protection of children from sexual exploitation
3.2.2.7	Promotional material offers true and complete information. No misleading information is being published
3.2.3	**Quality and sustainability criteria of customer satisfaction**
3.2.3.1	Customer communication raises awareness regarding sustainability aspects
3.2.3.2	After recertification: Customers are informed about the CSR label and the CSR process
3.2.3.3	Customers are informed about travelling with environmental friendly means of transportation
3.2.3.4	Minimum and maximum number of participants are listed
3.2.3.5	Customers are explicitly informed if the destination holds the risk of sexual exploitation of children
3.2.3.6	Customers receive information and interpretation about human rights and about the social, economic and cultural situation of the local population as well as about the natural surroundings
3.2.3.7	Customers are advised on appropriate behavior in the destination (taking pictures, tips, decent clothes, etc.).
3.2.3.8	Customers receive information about the protection of flora and fauna. No animals, plants or products protected by the Convention on International Trade in Endangered Species (CITES) or other laws are exported
3.2.3.9	Customers receive qualified recommendations about literature, guide books and information material
3.2.3.10	Customers receive information about appropriate behavior when participating in specific sports activities
3.2.3.11	Customers receive information about the protection of cultural heritage and the prohibition on the exportation of antiques and other objects of cultural value
3.2.3.12	Customers receive information about safety aspects in the destination with reference to information provided by the Foreign Office
4	**Employees**
4.1	**Human resources development**
4.1.1	All employees, including those being based in the destinations, receive CSR training (at least annually) comprising aspects about human rights, biodiversity, climate protection, working conditions and health and safety
4.1.2	Companies with more than 50 employees: training hours per employee / training hours per employee regarding human rights, biodiversity, climate protection, working conditions and health and safety
4.1.3	The enterprise pays its employees according to collective-agreement wages
4.1.4	The enterprise has overtime regulations
4.1.5	The enterprise has a system of voluntary social benefits
4.1.6	The enterprise supports staff representation/worker's council
4.2	**Employee structure**
4.2.1	Number of management positions, tourism professionals, trainees, other employees
4.2.2	Percentage of female employees in management positions

(continued)

Table 5.9 (continued)

4.2.3	Percentage of female employees
4.3	**Employee satisfaction**
4.3.1	Total number and percentage of participants of the survey
4.3.2	*The enterprise carries out an anonymous employee survey containing the following criteria. The survey results in a CSR index for employee satisfaction based on the questions of "strategy and management" and "working conditions"*
4.3.3	**Criteria/questions regarding strategy and management**
4.3.3.1	Do your direct seniors act in accordance with the basic values and objectives in the day-to-day business operations?
4.3.3.2	Do you, in your daily work, use the company's philosophy and policy as an orientation?
4.3.3.3	Are the management decisions clear and understandable to you?
4.3.3.4	Can you, within your field of competence, act and decide independently?
4.3.3.5	Are the tasks in your organisation distributed in an appropriate manner?
4.3.3.6	Are operations and procedures properly defined in your field of work?
4.3.3.7	How do you estimate customer satisfaction?
4.3.3.8	Does your organisation value the opinion of employees, and can you contribute your own ideas and objectives?
4.3.3.9	How do you estimate the profile and image of your organisation?
4.3.3.10	Do you think the travel offers sold by your company contribute to sustainable development?
4.3.3.11	How important is the sustainability commitment of your company to you?
4.3.3.12	Do you identify yourself with your company?
4.3.4	**Criteria/questions regarding working conditions**
4.3.4.1	How satisfied are you with the opportunities for continued training and capacity building?
4.3.4.2	Does your organisation offer you possibilities for job development and promotion?
4.3.4.3	To what extent can you make use of your skills in your field of work?
4.3.4.4	Does your organisation offer a flexible job design and working hours?
4.3.4.5	Is your remuneration adequate for the work that you do?
4.3.4.6	Are health and safety standards met at your workplace?
4.3.4.7	Does your organisation offer family-friendly working conditions?
4.3.4.8	Does your organisation realise gender equality and equal rights of people from different age groups and people from different cultural backgrounds?
4.3.4.9	Is the working atmosphere characterised by constructive cooperation and mutual support?
4.3.4.10	Do you feel that your job is secure on the long term?
4.3.4.11	Do you enjoy your work, all in all?
4.3.5	**Criteria/questions regarding environmental protection in the office**
4.3.5.1	How do you rate environmental protection activities in your organisation in the following fields? – Heating and warm water – Electricity – Waste – Water consumption – Procurement (paper, material, etc.)

(continued)

Table 5.9 (continued)

	Consumable goods (Fair Trade, regional, organic)
	– Business travel
4.3.5.2	How do you rate your own behavior with regard to environmental protection?
5	**Business ecology**
5.1	**Energy**
5.1.1	The enterprise uses 100 % green power from renewable sources
5.1.2	Power consumption in total and per employee
5.1.3	Heat consumption in total and per employee
5.2	**Emissions**
5.2.1	Transportation to work: CO_2/km per means of transport
5.2.2	Business trips: CO_2/km per means of transport
5.2.3	CO_2 emissions in total and per consumer (electricity/heat/transportation, etc.)
5.2.4	CO_2 compensation of business trips (at least 50 %)
5.2.5	CO_2 emissions per employee
5.3	**Paper**
5.3.1	Weight of paper per employee
5.3.2	Weight of catalogue paper per customer
5.3.3	Percentage of recycled paper
5.4	**Procurement**
5.4.1	The enterprise has a guideline regarding fair trade/regional/organic/etc. procurement (consumable goods, office supplies, events, etc.)
6	**Product responsibility in the supply chain**
6.1	**Partner agencies**
6.1.1	Total number of partner agencies
6.1.2	Number of checked partner agencies
6.1.3	*The enterprise checks and evaluates its partner agencies against the following criteria. The evaluation results in a CSR index for partner agencies*
6.1.4	**Criteria regarding management of the partner agencies**
6.1.4.1	Does the enterprise have a contract with the partner agency?
6.1.4.2	Is the partner agency familiar with criteria and principles of sustainable tourism?
6.1.4.3	Does the partner agency have a quality certificate? If yes please name the certification
6.1.4.4	Does the partner agency have an environmental certificate? If yes please name the certification
6.1.4.5	Has the partner agency signed the Code of Conduct for the protection of children from sexual exploitation?
6.1.4.6	Are there any other signed voluntary commitments (please name if applicable)?
6.1.4.7	Does the partner agency actively prevent exploitative child labor within its immediate infrastructure?
6.1.4.8	Does the partner agency ensure the safety of its employees?
6.1.4.9	Does the partner agency comply with national regulations in the labor legislation?
6.1.4.10	Does the partner agency pay fair and appropriate wages to its employees but at least the minimum wage?
6.1.4.11	Does the partner agency systematically evaluate and improve the customer satisfaction (possibly together with tour operator)?

(continued)

Table 5.9 (continued)

6.1.4.12	Does the partner agency provide training for employees and create long-term work perspectives?
6.1.5	**Ecological criteria: the partner agency:**
6.1.5.1	Acts ecologically within own infrastructure
6.1.5.2	Actively seeks ways to reduce waste generated by consumable goods in the office and offered to guests
6.1.5.3	Considers environmentally friendly behavior when subcontracting further suppliers
6.1.5.4	Prefers certified suppliers when subcontracting
6.1.5.5	Promotes environmentally friendly behavior of suppliers by providing training or awareness-raising measures
6.1.6	**Socio-economic criteria—the partner agency**:
6.1.6.1	Is socio-cultural sensitive
6.1.6.2	Prefers socio-cultural suppliers when subcontracting further suppliers
6.1.6.3	Prevents the use of tourist infrastructure which deprives local communities of scarce resources
6.1.6.4	Maintains an open and fair relationship with its business partners
6.1.6.5	Chooses local suppliers when awarding subcontracts
6.1.6.6	Pays its subcontractors appropriately and fair
6.1.6.7	Ensures the safety of its customers
6.2	**Accommodations**
6.2.1	Total number of accommodations
6.2.2	Number of checked accommodations
6.2.3	Number of accommodations holding a recognised sustainability certificate
6.2.4	The enterprise checks and evaluates its accommodations against the following criteria. The evaluation results in a CSR index for accommodations
6.2.5	**Criteria regarding management of the accommodation**
6.2.5.1	Does the enterprise have a contract with the accommodation?
6.2.5.2	Does the accommodation have a quality certificate? If yes please name the certification
6.2.5.3	Does the accommodation have an environmental certificate? If yes please name the certification
6.2.5.4	Has the accommodation signed the Code of Conduct for the protection of children from sexual exploitation?
6.2.5.5	Are there any other signed voluntary commitments (please name if applicable)?
6.2.5.6	Does the accommodation actively prevent exploitative child labour within its immediate infrastructure?
6.2.5.7	Does the accommodation ensure the safety of its employees?
6.2.5.8	Does the accommodation comply with national regulations in the labour legislation?
6.2.5.9	Does the accommodation pay fair and appropriate wages to its employees but at least the minimum wage?
6.2.5.10	Does the accommodation systematically evaluate and improve the customer satisfaction (possibly together with tour operator)?
6.2.6	**Ecological criteria—the accomodation:**
6.2.6.1	Ensures that its activities do not endanger the neighbouring communities' basic need of food, water, electricity and sanitation
6.2.6.2	Informs its guests about ecological behaviour

(continued)

Table 5.9 (continued)

6.2.6.3	Offers organic and/or regional products
6.2.6.4	Does not harvest, consume, display or trade wildlife species, except as part of a regulated activity that ensures that their utilisation is sustainable and in compliance with local to international laws
6.2.6.5	Does not hold wildlife except for properly regulated activities in compliance with local to international law
6.2.6.6	Only keeps protected and wildlife species when hold by persons authorised and suitably equipped to house and care for them humanely
6.2.6.7	Has implemented a low-waste purchasing policy
6.2.6.8	Uses environmental friendly detergents
6.2.6.9	Reuses wastewater wherever possible. Other waste water is being disposed of in an environmentally friendly way, so that there is no harm for neighbouring communities
6.2.6.10	Uses renewable energies
6.2.6.11	Knows, who provides its drinking water supply and systematically evaluates and reduces its water consumption
6.2.6.12	Separates its waste and systematically evaluates and reduces non-recyclable waste
6.2.6.13	Ensures that its waste disposal has no negative impact on the environment and on neighbouring communities
6.2.6.14	Actively promotes energy saving measures
6.2.6.15	Reduces or replaces pollutants (pesticides, paints, swimming pool disinfectants, detergents). All harmful substances are stored, used and disposed of adequately
6.2.6.16	Implements strategies to minimise pollution from noise, light, runoff, erosion, greenhouse gases, and air, water and soil contaminants
6.2.7	**Socio-economic criteria—the accomodation:**
6.2.7.1	Complies with local architectural standards
6.2.7.2	Is barrier-free and accessible for people with special needs
6.2.7.3	Employs local staff
6.2.7.4	Employs local staff in management positions
6.2.7.5	Provides regional cuisine
6.2.7.6	Trains its employees on sustainability aspects on a regular basis
6.2.7.7	Prevents the use of tourist infrastructure which deprives local communities of scarce resources
6.2.7.8	Promotes the protection of historically, archaeologically and spiritually important places. Tourist offers do not impede access by local
6.3	**Tour guides**
6.3.1	Total number of tour guides
6.3.2	Number of checked tour guides
6.3.3	*The enterprise checks and evaluates its tour guides against the following criteria. The evaluation results in a CSR index for tour guides.*
6.3.4	**Criteria regarding the quality of cooperation with the tour guide**
6.3.4.1	Does the enterprise have a contract with the tour guide?
6.3.4.2	Does the tour guide have an official authorisation? If yes please name the authorisation
6.3.4.3	Does the tour guide receive training on a regular basis (if applicable please indicate date of last training)
6.3.4.4	Has the tour guide been informed about the tour operator's mission statement?

(continued)

Table 5.9 (continued)

6.3.4.5	Is the tour guide qualified to assess the security situation, to take preventive actions and to provide first aid?
6.3.4.6	Is the tour guide committed to and actively communicates child protection concerning sexual exploitation?
6.3.4.7	Does the tour guide receive a fair and appropriate salary?
6.3.4.8	Does the tour guide speak the locally used language?
6.3.5	**Ecological criteria—the tour guide:**
6.3.5.1	acts exemplary in an environmental friendly way
6.3.5.2	informs his/her guests about the protection of flora and fauna
6.3.6	**Socio-economic criteria—the tour guide:**
6.3.6.1	is adequately knowledgeable about the region
6.3.6.2	behaves in a sensitive and respectful manner when interacting with the local population
6.3.6.3	treats cultural heritage respectfully and provides information on correct behaviour when visiting cultural heritage sites
6.3.6.4	raises awareness regarding local added value
6.3.6.5	provides guests with information regarding adequate behaviour in host country (tipping, clothing, taking photos, etc.)
6.3.6.6	provides guests with information about safety standards and regulations
7	**Community Involvement**
7.1	The enterprise has defined criteria for choosing sustainability projects (environment, socio-cultural, education, health, etc.)
7.2	Average duration of projects and partnerships

analysed with regard to sustainability aspects. The data gathered will be assessed and suggestions for improvement in the various areas will be collected and priorities will be defined.

- Setting up a programme for improvement: strengths and weaknesses will emerge from the assessment of the current situation. In the policy, the direction has been laid down in which the company is to move in the future. A programme for improvement with clear and measurable targets, including measures to achieve them, shows in a transparent manner what is to be achieved next. Not all the identified potentials for improvements can be addressed at once. It is advisable to develop the first programme for improvement covering a time frame of 2–3 years. Progress is to be assessed annually, using indicators to determine to what extent the objectives have been achieved, and the programme is to be updated annually. The programme for improvement needs to be agreed upon and adopted by the top management.
- Writing a CSR report: every year a CSR report has to be prepared. How the report is to be illustrated and whether it is to be printed or not depends on the ways of communication to be used to make it wide accessible. Then it has to be published through all communication channels available (e.g. catalogue, website, press)
- External auditing: the report can be audited by a specialised external organisation. If the criteria are met, a CSR label (CSR report certified) can be issued.

TourCert is the certification system adopted (www.tourcert.org) and the certification process is ratified by a network of independent European experts working for universities and international bodies.

It is evident that the CSR Reporting Standards requires firstly a self-assessment by the tour operator and secondly the audit by an external and independent third-party (TourCert) that grants the certification with the logo "CSR Tourism Certified".

Summarising the main characteristics of the system, the CSR Reporting Standards is based on the certification approach, combining process and performance standards. The responsible tourism perspective is very marginal, in that the main goal is to assess the responsibility within the tour operator and along the supply chain, including providers and distributors as partners, accommodations, tour guides and travel agencies. Indicators do not verify whether tour operators respect the principles of responsible tourism.

Looking at the triple bottom line, although the indicators cover all areas of sustainability, the social aspect seems to be less highlighted than environmental and economic dimensions. Although great attention is given to the level of responsibility of the company towards clients and staff, there are no indicators regarding the involvement of local community and the organisation of meetings between tourists and local people during the stay. For example, tour operators could be required to prove that the financed actions are taken in agreement with the local community. Or there are no indications about whether the decisions taken about the travel planning and organisation involve representatives of local stakeholders, or that tourist guides provide tourists information on local culture, history and customs. An explanation stays probably in the difficulty to translate social aspects into measurable and objective indicators, as the other parameters developed by KATE. AITR and ATES use qualitative indicators, asking the operator to enclose documents proving the accomplishment of some standards.

As for the application to tour operators of different size, although the CSR Reporting Standards can be applied by all companies, regardless of their size, this procedure seems more suitable for medium than for small and micro tour operators, because it requires the collection of a large amount of data and the development of a number of indicators, which can be complex for operators with a small staff of people. However, the successful adoption of the system by several tour operators (about 70 companies) seems to show its feasibility. In addition, the quality of the CSR Reporting Standards seems to be confirmed by its application by other companies outside Germany (for example in Italy) and by the development of a parallel CSR system for accommodations.

Fig. 5.5 ATR's Logo

5.6 ATT: Agir Pour un Tourism Responsable

5.6.1 Organisation and Objectives

ATT (Association des Tours Opérateurs Thématiques, in English Association of Thematic Tour Operators) was created in France in 2004 and is formed by small- and medium-size tour operators. They organise travels "with a sense", generally following a specific theme, and promote responsible behaviour among operators and tourists, so as to respect the social, cultural and environmental features of the destination.

ATT has developed the ATR reporting system (ATR—Agir pour un Tourisme Responsable, in English Acting for Responsible Tourism), addressed to its members, with the aim of making tour operators and other tourism stakeholders aware of their influence and impacts on the destination and involve them in a responsible activity (from travel planning to the concrete development of the travel package and to the real experience lived by tourist).

The main objectives of the system are as follows:

- Encourage the development of qualitative tourism, so as to ensure a true discovery of destinations;
- Ensure that the economic impacts are equally distributed among all stakeholders involved in the creation of the travel package at local level;
- Respect local populations and involve them in local development, favouring the direct employment of local people on equal expertise and the use of local resources to build the travel package;
- Respect local customs and different cultures at the destinations;
- Respect the environment and the local natural and cultural heritage;
- Inform tourists sincerely and transparently;
- Apply the principle of responsible management also within their enterprise (Fig. 5.5)

5.6.2 The Reporting System

Like ATES, also ATT asks tour operators who join the program to present a list of technical documents in order to prove the respect of the association's principles (Table 5.10).

Table 5.10 ATT. Technical documents required by ATR system

1. Sales document
2. Description of communication tools
3. ATR Code of Ethics
4. Technical Sheet by country or region, providing information on the economic, social and environmental background in the destination, on the behaviour to be respected and other information
5. Sheet about the local staff
6. Training plan provided to local providers
7. Questionnaire for the customer satisfaction
8. Travel plan for every travel group, providing the number of participants and the program
9. Contracts
10. Documents that prove the transfer of information to ATT by the company
11. Providers' licence
12. Destination sheet
13. Contract between the tour operator and the providers
14. Training plan provided to employees
15. Document that prove that the tour operator use the same providers for at least 3 years
16. Sheet describing social benefit for employees
17. Description of projects financed by the tour operator
18. Environmental policy
19. Internal communication about the company's development
20. Description of ATT project in which the tour operator takes part.

The reporting system refers to the level of responsibility of the tour operator in doing its activity. It includes four main commitments, divided into a number of criteria, which are explained through 24 indicators. These indicators in particular represent specific actions that tour operators have to carry out and prove through proper documentation (Table 5.11).

To each indicator a numerical value is associated, so as to determine the tour operator level of achievement of each objective according to a range already established (minimum and maximum values).

In order to obtain the certification, as the CSR reporting Standards of KATE, the ATR system requires a self-assessment by the company and an evaluation carried out by an external and independent third party. The body chosen by ATT for granting the certification is AFNOR Certification, an organisation acknowledged at international level for the assessment and certification of products and services.

At the moment, 13 associates have awarded the certification and other two members are completing the auditing procedure.

Summarising the main characteristics of the ATR system, although it basically follows a process-based certification approach, it partially includes a responsible tourism perspective. The certification process gives great attention both to clients and to their opinions and perceptions, and to a proper public communication by the TO of the commitments he is taking according to the principles of responsible tourism. The ATR scheme also gives importance to community involvement and highlights the relevance of developing lasting relationships with local stakeholders

Table 5.11 ATT. Criteria and indicators of ATR system

		Documents
1	**Provide clients with clear information and make them aware of responsible tourism**	
1.1	**Provide a general information to clients**	
1.1.1	The client receives clear and readable information about the proposed services and the commitment of tour operator to responsible tourism	
1.1.1.1	The TO describes precisely in its communication materials (quotation and/or technical documents) the range of services included in the travel package	1, 2
1.1.1.2	The TO shows clearly his commitment to responsible tourism and its principles on his communication materials	2
1.2	**Make clients aware of responsible tourism**	
1.2.1	**The respect of local populations and cultures**	
1.2.1.1	The TO distributes to his clients the ATR Code of Ethics	3
1.2.1.2	Whenever needed (the geographical distance of the destination often requires specific information to clients), the TO provides to clients information on the destination proposed	4
1.2.2	**The clear communication of actions taken by the TO for the preservation of the environment and/or local development**	
1.2.2.1	The TO provides clear communication (also with detailed figures), on his main communication tools, about the actions carried out for the preservation of the environment and the local development according to the ATR principles	2
1.2.3	The proposal to take part in one or a number of initiatives for the preservation of the environment and/or local development and/or the diffusion of cultures, related to tourism	
1.2.3.1	The TO proposes to his clients to take part to the previous actions, both launched according to personal or ATR initiative	20
2	**Bring clients at destination and guide them according to the principles of responsible tourism**	
2.1	**A selection of local stakeholders/partners who support local employment**	
2.1.1	At least 60 % of travellers are accompanied by local staff. This commitment is not applicable to independent travel package	5
2.1.2	At least 60 % of travellers are lodged in local accommodation establishments, whose staff is for 75 % coming from the area	5
2.2	**The transfer of competence to identified local stakeholders**	
2.2.1	After the assessment of needs, the TO applies, and check the presence of, a training program for accommodation establishments and local tourist guides (Tour leaders)	6
2.3	**Inform clients to respect local environment, heritage and cultures**	
2.3.1	During the stay, the Tour leader informs clients and make them aware on the local environment, the heritage and local cultures	7

(continued)

Table 5.11 (continued)

		Documents
2.4	**Management of tourist flows by destination**	
2.4.1	The TO limits the travel groups to a maximum of 15 people for adventure travels and to 25 people for other type of travels	8
2.4.2	The TOs organising seminars, cruises, event travels and incentive travels that involve a large number of travellers, organise these travels so as to avoid a congestion of destinations and use proper resources (financial and human) for the cleaning and restoration of sites	9
2.4.3	The TO commits to communicate to ATR the degradation of the visit conditions of a destination so as the network push local authorities and tourism stakeholders to develop a proper management policy of tourist flows	7, 10
3	**Identify the stakeholders/partners according to the principles of responsible tourism**	
3.1	**The stakeholders/partners run their business according to the legal rules in force in their countries**	
3.1.1	The TO only works with stakeholders and partners respecting the legal rules and specifically the tourist regulation in force in their countries, wherever they are present	11
3.2	**The selected stakeholders/partners assure proper work conditions to their employees, taking into account specific local conditions**	
3.2.1	The TO respects the rules established by ATR and concerning the work conditions of local stakeholders, according to the destinations. If a destination is not mentioned in these rules, the criterion is not applicable	7, 12
3.3	**The selected stakeholders/partners are made aware towards the application of a proper waste management policy and natural resource management policy**	
3.3.1	For open-air trips, including short excursions, the policy concerning the management of waste and the use of natural resources established by ATR is communicated by the TO to local stakeholders with contract documents	13
3.3.2	For trips involving hotel stays, a document prepared by ATR and stimulating a responsible policy by the hoteliers is distributed by the TO with the contract documents	-
3.4	**The TO commits himself to a long-lasting relationship with its stakeholders/partners**	
3.4.1	The TO states that he favours lasting relationships with the stakeholders and commits himself to not terminate a contract illegally	13
3.4.2	The TO states that there are no annual changes of stakeholders in the same destination during 3 years in a row, except for specific situations	15
4	**The TOs commits himself to apply the principle of responsible management also within their enterprise**	
4.1	**Training the internal staff according to the principles of responsible tourism**	
4.1.1	The staff training plan includes (for staff involved in the organisation and selling of travels and for tour leaders) a set of actions to make them aware	14

(continued)

Table 5.11 (continued)

		Documents
	of the importance to respect local populations and cultures and to preserve environment and local heritage	
4.2	**Applying an environmental management policy within the enterprise**	
4.2.1	The TO develops and applies within its enterprise a proper environmental policy	15
4.3	**Managing the communication policy**	
4.3.1	The TO provides staff regularly with information about the enterprise and its development	7
4.4	**Applying a procedure to share the value created by the activities**	
4.4.1	The TO develops a system of social benefits to share the value created by the enterprise's activities with the staff	18
4.5	**Supporting actively and financially the development of local populations**	
4.5.1	The TO takes part in actions proposed by ATR	19
4.6	**Taking into account the clients' satisfaction deriving from the enterprise's commitments**	
4.6.1	The TO includes in the customer satisfaction questionnaire some questions regarding the social and environmental aspects proposed by ATR	16

as well as of preserving the environment and respecting different cultures; however there are no specific indicators about the meeting between local population and tourists.

The system considers as important stakeholders not only the consumers and the community, as the schemes of AITR and ATES, but also the providers and the employees, as the programs of KATE-TourCert and QUIDAMTUR.

Considering the triple bottom line, the ATR approach focuses above all on social and environmental aspects. The economic dimension is less developed, since there are not indicators related to the company economic performance or to the share of the package price paid by clients that go to the advantage of local population.

It is interesting to mention the importance given to training activities the company should develop for the employees and local stakeholders/partners, so as to make them aware of responsible tourism and push them to adopt a proper behaviour. The transfer of skills, competences and best practices can be very important to support the development of destinations and to assist local enterprises in developing countries.

5.7 TOI-GRI: Tour Operator Sector Supplement

5.7.1 Organisation and Objectives

TOI, the Tour Operators Initiative for Sustainable Tourism Development, is a network of tour operators from many different countries that have committed themselves to integrate sustainable development into their business practices and to work together through common activities in order to promote and disseminate methods and practices compatible with sustainable development. The Initiative, registered in Switzerland in 200, is voluntary, non-profit and open to all tour operators, regardless of their size and geographical location.

The Initiative has been developed by tour operators for tour operators with the support of the United Nations Environment Programme (UNEP), the United Nations Educational, Scientific and Cultural Organization (UNESCO) and the World Tourism Organization (UNWTO), who are also full members.

Under this international umbrella, tour operators who are members of the Initiative should be able to respond to international agendas while creating a platform to develop ideas and projects to address the environmental, social, economic and cultural aspects of sustainable development within the tourism sector.

The mission of TOI Initiative is:

- To advance the sustainable development and management of tourism;
- To encourage tour operators to make a corporate commitment to sustainable development.

This mission should be achieved by pursuing the following strategic objectives and associated actions.

- Assist tour operators in implementing sustainable tourism through the adoption of responsible practices in their internal operations and management of the supply chain, and in destinations. This objective can be achieved by facilitating the exchange of best practices, developing new management tools and adapting existing ones, implementing programmes for sustainable development in destination, involving tour operators in the relevant programmes of work of UNEP, UNESCO, UNWTO, assessing progress made on a regular basis, facilitating partnerships between members to address issues with a common voice.
- Broaden the support for sustainable development among other players in the tourism sector, including tourists, by cooperating with key stakeholders, increasing the awareness of key players, like tourists, other segments of the tourism industry, local communities and people, tourism trade associations and local and national authorities.
- Create the critical mass of committed tour operators by increasing their visibility, establishing partnerships with other organisations that contribute to the achievement of the same objectives, strengthening links with regions through tour operators' associations and UNEP, UNESCO and UNWTO networks.

Among all activities, TOI has developed a methodology for tour operators seeking to integrate economic, environmental and social sustainability criteria into their activity and for the choices of service suppliers. This methodology includes three key components.

- Establish a sustainable supply chain policy and management system, based on a sustainability action plan for improving the economic, environmental and social sustainability performance of suppliers; a baseline assessment of tourism service suppliers; a system for monitoring and reporting on progress.
- Support suppliers in reaching sustainability goals by raising awareness on sustainability issues, demonstrating why sustainability performance is important, providing technical support to suppliers on how to improve their sustainability performance; offering incentives to sustainable suppliers by recognising and rewarding improvements on key environmental, social and economic issues.
- Integrate sustainability criteria into suppliers' contracts, setting minimum performance standards that all suppliers of a particular type must meet and further optional criteria that they are encouraged to achieve.

In 2002, TOI also co-operated with GRI—Global Reporting Initiative—to develop a sustainability reporting framework. GRI is an international network-based organisation that has developed one of the world's most widely used sustainability reporting framework—not only in tourism.

In order to ensure the highest degree of technical quality, credibility, and relevance, the reporting framework developed by GRI is developed through a consensus-seeking process with participants drawn globally from business, civil society, labour and professional institutions. This framework sets out the principles and indicators that companies can use to measure and report their economic, environmental and social performance.

It is formed in particular by three main parts.

- The Sustainability Reporting Guidelines, which contain the basic principles and are divided into three main areas (economic, social and environmental). The guidelines include the general information and the general list of indicators to be respected by every company, regardless of the industry sector it belongs to.
- The Sector Supplement, which integrates the Guidelines providing a set of specific indicators for each industry sector; although it is important that a reporting system is based on a single set of indicators in order to facilitate comparisons among businesses, it is also necessary to integrate appropriate sustainability indicators that are able to assess the specificity of every industry sector.
- The Technical Protocols, which offer practical and technical advices to users.

The cooperation between TOI and GRI has resulted in particular in the development of the pilot version of the Tour Operators' Sector Supplement, thanks to a multi-stakeholder working group formed by members of TOI and GRI. The scope was to capture issues that are either essential components of sustainability specific of tour operators or relevant to numerous sectors, but of critical importance to tour operators' sustainability performances. Therefore this supplement only applies to businesses dedicated to organising holiday packages (Figs. 5.6 and 5.7).

Fig. 5.6 TOI's Logo

Fig. 5.7 GRI's Logo

5.7.2 The Reporting System

The Tour Operators' Sectors Supplement requests information on tour operators' performance in terms of policies and management systems as well as of areas of impact. The system is based on five categories, with in total 126 indicators that reflect the life cycle of the holiday product: from the planning stage, to the development and delivery of the product.

- Product management and development (PMD): it includes actions related to the choice of the destination as well as the type of services to be included.
- Internal management (IM): it reflects all the operations and activities that take place in the headquarters or country offices;
- Supply chain management (SCM): it addresses actions related to the selection and contracting of service providers;
- Customer relations (CR): it summarises the actions taken to deal with consumers, not only with regards to the responsibility to serve them and reply to their comments but also to the opportunity to provide information and raise consumer awareness regarding sustainability;
- Co-operation with destination (D): it includes all activities and decisions related to destinations that tour operators make beyond the production and delivery of their holiday package. This mainly includes efforts made by tour operators to engage in dialogues with local operators about the impacts of tour packages and philanthropic activities.

While the first four categories cover actions and decisions that tour operators make in preparation of the holiday package and that have either a direct impact or a significant influence on sustainability, the fifth area covers efforts to address product impacts at the destination.

Table 5.12 describes all indicators of the Tour Operators' Sector Supplement. It has to be noted that for some indicators there is a partial overlap with those included in the Guidelines, since the Supplement is a specification of the more general Guidelines.

Table 5.12 TOI-GRI. Indicators of the tour operators' sector supplement

Product management and development (PMD)	
PMD 1	Indicate % of the business activity (by passengers carried) and market share in each operating destination
PMD 2	Describe policies on selecting, developing and deselecting destinations based on environmental, social and economic issues
PMD 3	Describe key environmental, economic and social issues identified in destinations and types of information gathered by the company Issues may include: *For the environment: water, wastewater, energy, and transport infrastructures, hazardous and solid waste disposal, air and water quality land-use and biodiversity conservation, local environmental management structures* *For the socio-economic dimension: respect for human rights, compliance with International Labour Organization (ILO), conventions on child and forced labour, freedom of association, non-discrimination, programmes to combat sexual exploitation of children and to combat and mitigate the social impacts of HIV/AIDS, citizen participation in planning process*
PMD 4	Indicate % of destinations in which organisation operates for which issues (PMD3) have been identified, and percentage of business activity in these destinations (by passengers carried)
PMD 5	Describe and motivate types of approaches taken in gathering information (PMD3) in a given destination *Approaches include own research, third party report, customer feedback, destination self-assessment, stakeholders' consultation, dialogue with competitors. Specify stakeholders approached*
PMD 6	Indicate length of time over which this information (PMD3) has been collected, and the frequency with which it is updated
PMD 7	Quantify overall economic, environmental, and social impacts of typical holiday products
PMD 8	Describe changes in design of holiday packages and other actions to address key environmental, economic and social issues (see PMD3) in destinations
PMD 9	Describe measures adopted to maximise economic benefits to destinations
Internal management (IM)	
IM 1	Provide evidence of recruiting local residents (including destination nationals) for destination posts including management positions
IM 2	Describe existence of policies and programmes to address the physical and mental well-being of staff at headquarters and destinations
IM 3	Describe types and mechanisms of training on environmental, social and economic issues by category of employee. Include destination representatives, senior management, middle management, professional, technical, administrative, production, and maintenance
IM 4	Describe policies and actions in place to accommodate cultural customs, traditions and practices of staff throughout the organisation
IM 5	Describe policies to minimise the environmental impacts associated with the production, distribution and use of promotional materials and customer documentation
IM 6	Indicate total quantity (tones or kg) of material used by type (e.g., paper, plastic) and environmental quality (e.g., recycled content), for the production of promotional materials and customer documentation
IM 7	Indicate percentage of promotional materials and customer documentation that are produced in accordance with an environmental standard. Specify standard
IM 8	Describe policies and targets for redistribution, reuse and recycling of promotional materials

(continued)

Table 5.12 (continued)

IM 9	Indicate percentage of total travel retailers that agree to adopt policies and practices on reuse and recycling of promotional materials. Also refer to Supply Chain Management section
Supply chain management (SCM)	
SCM 1	Describe the supply chain management policy, objectives and targets on environmental, social, and economic performance
SCM 2	State the use of supplier prioritisation and screening criteria
SCM 3	Describe processes through which suppliers, by type, are consulted during development and implementation of the supply chain management policy, described in SCM 1
SCM 4	Describe issues identified through supplier consultation and actions to address them *Processes include: One way communication (*e.g., *questionnaires), two-way communication (*e.g., *information exchange), active co-operation (*e.g., *supplier training), rewards and recognition for high performers*
SCM 5	State joint actions taken with suppliers, by type, to support improvements in suppliers own environmental and social performance
SCM 6	Describe progress in achieving objectives and targets related to supply chain policy
SCM 7	Indicate percentage of suppliers, by type, subject to supply chain management policy
SCM 8	Indicate percentages of suppliers, by type, subject to supply chain policy that have a published sustainability policy, implemented a sustainability management system and/or have a staff person with management responsibility for corporate sustainability
SCM 9	State types of information requested from suppliers, by type, on their: (a) Environmental practices and performance *Include: Materials, water, energy, purchasing, solid waste, hazardous waste, effluents, emissions, transport, land-use and biodiversity* (b) Social practices and performance *Include: Community and staff development, indigenous and tribal people's rights, formal employment contracts, social security, working conditions according to ILO Convention 172, equal treatment, non-discrimination, recognition of independent trade unions and application of collective bargaining agreements, health and safety committees, policies excluding child labour as defined by ILO, programmes to combat commercial sexual exploitation of children and to combat and mitigate the social impacts of HIV/AIDS*
SCM 10	Indicate percentage of suppliers, by type, subject to supply chain management policy that provided the requested information
SCM 11	Indicate percentage of suppliers, by type, subject to supply chain management policy whose environmental, social and economic performance has been reported *Through, for example: supplier self declaration; spot checks by reporting organization; environmental and social audits; certification schemes (including eco labels); third party verification (state if verifier is accredited, and by whom)*
SCM 12	State actions taken by the reporting organisation in response to suppliers reported performance (as per SCM 11), by type of suppliers. Include incentives and rewards
SCM 13	State actions to inform suppliers of customers' requirements
SCM 14	State contracting policy and how it is communicated to suppliers. Include negotiating terms and conditions for payment, cancellation and compensation of contracts with suppliers
SCM 15	Describe joint initiatives with suppliers to improve environmental, social and economic conditions in destinations

(continued)

Table 5.12 (continued)

SCM 16	State benefits for the reporting organisation from implementing the sustainable supply chain policy
Customer relations (CR)	
CR 1	Describe tools and measures used by reporting organisation to raise the awareness of consumers of suppliers' environmental, social and economic performance *Indicate joint efforts with suppliers, NGOs, national tourism organisations (NTOs), and authorities in destinations; main themes communicated; and specify the moment of the holiday experience, from selection to return, at which communication took place*
CR 2	Describe tools and measures used by reporting organisation to raise the awareness of consumers of destinations' environmental, social and economic issues. *Indicate joint efforts with suppliers, NGOs, NTOs, and authorities in destinations; main themes communicated; and specify the moment of the holiday experience, from selection to return, at which communication took place*
CR 3	Describe tools and measures used by reporting organisation to raise the awareness of consumers of sustainable holiday-making *Indicate joint efforts with suppliers, NGOs, NTOs, and authorities in destinations; main themes communicated; and specify the moment of the holiday experience, from selection to return, at which communication took place*
CR 4	Describe means to invite customers' feedback on economic, environmental and social issues related to the holiday product and actions taken to respond to feedback
CR 5	Indicate percentage of total feedback received, related to economic, environmental and social issues *Break down feedback in terms of positive and negative and main issues covered*
CR 6	Provide evidence of consultation with destination stakeholders and suppliers on how the destination and services are portrayed to customers
CR 7	Indicate number of complaints from destinations' stakeholders and holiday-makers regarding misleading and inaccurate representation of destinations *Indicate actions taken to address complaints*
Cooperation with destinations (D)	
D 1	Describe ways reporting organisation engages with destination stakeholders to address issues, including those identified in PMD 3 *Include information exchange, awareness raising, consultation, dialogues and incentives provided.*
D 2	Describe measures taken to identify and offer commercial opportunities and assistance to non-contracted suppliers that support community development
D 3	Describe procedures and criteria for selecting projects and organisations to which philanthropic and charitable donations are made
D 4	Indicate total funds (in cash and estimated value of in-kind contributions) for conservation and social development projects *Identify: sources (corporate budget, consumers' contributions), type (cash, in-kind), distribution of funds to conservation and community development projects and organisations, location of projects/organisations receiving the funds (in operating destinations, headquarters and international).*
D 5	Describe programmes for philanthropic and charitable donations in relation to conservation and community development projects
D 6	Provide evidence of benefits generated (in D 4 and D 5), particularly at destinations, in support of community development, biodiversity conservation and other social, economic and environmental improvements at destinations

Other important indicators to determine the enterprise responsibility that are not included in the Sector Supplement are listed in the General Guidelines that are common to all sectors. They refer to labour, human rights, society and product responsibility. Among them, it is particularly important to mention:

- EC 1 Direct economic value generated;
- EC 5 Range of ratios of standard entry level wage compared to local minimum wage at significant locations of operation;
- EC 6 Policy, practices, and proportion of spending on locally-based suppliers at significant locations of operation;
- EC 7 Procedures for local hiring and proportion of senior management hired from the local community at significant locations of operation;
- EC 8 Development and impact of infrastructure investments and services provided primarily for public benefit through commercial, in-kind, or pro bono engagement;
- EC 9 Understanding and describing significant indirect economic impacts, including the extent of impacts;
- EN 2 Percentage of materials used that are recycled input materials;
- EN 3, EN 8 Direct energy, water and other resource consumption;
- EN 5 Energy saved due to conservation and efficiency improvements;
- EN 7 Initiatives to reduce indirect energy consumption and reductions achieved;
- EN 11 Location and size of land owned, leased, managed in, or adjacent to, protected areas and areas of high biodiversity value outside protected areas;
- EN 12 Description of significant impacts of activities, products, and services on biodiversity in protected areas and areas of high biodiversity value outside protected areas;
- EN 14 Strategies, current actions, and future plans for managing impacts on biodiversity;
- EN 16 Total direct and indirect greenhouse gas emissions by weight;
- EN 18 Initiatives to reduce greenhouse gas emissions and reductions achieved;
- EN 26 Initiatives to mitigate environmental impacts of products and services, and extent of impact mitigation;
- EN 29 Significant environmental impacts of transporting products and other goods and materials used for the organisation's operations, and transporting members of the workforce;
- EN 30 Total environmental protection expenditures and investments by type;
- LA 1 Total workforce by employment type, employment contract and region;
- LA 2 Total number and rate of employee turnover by age group, gender and region;
- LA 3 Benefits provided to full-time employees that are not provided to temporary or part-time employees, by major operations;
- LA 7 Rates of injury, occupational diseases, lost days, and absenteeism and total number of work-related fatalities by region;

- LA 8 Education, training, counselling, prevention, and risk-control programs in place to assist workforce members, their families or community members regarding serious diseases;
- LA 10 Average hours of training per year per employee by employee category;
- LA 11 Programs for skills management and lifelong learning that support the continued employability of employees and assist them in managing career endings;
- LA 14 Ratio of basic salary of men to women by employee category;
- HR 2 Percentage of significant suppliers and contractors that have undergone screening on human rights and actions taken;
- HR 6 Operations identified as having significant risk for incidents of child labour and measures taken to contribute to the elimination of child labour;
- SO 1 Nature, scope, and effectiveness of any programs and practices that assess and manage the impacts of operations on communities, including entering, operating and exiting;
- PR 1 Life cycle stages in which health and safety impacts of products and services are assessed for improvement, and percentage of significant products and services categories subject to such procedures;
- PR 5 Practices related to customer satisfaction, including results of surveys measuring customer satisfaction.

These indicators are only a part of the set of 80 indicators included in the Guidelines that testify that the GRI Reporting Framework is particularly complete. Indeed all three dimensions of the triple bottom line are covered with appropriate indicators. It has to be noted that there are specific references also to the relationships with the local community that is neglected in other methodologies. The tour operator is required to prove these relationships (indicators D 2 and EC 6), to provide evidence of consultation with destination stakeholders and suppliers on how the destination and services are portrayed to customers (indicator CR 6) and to describe ways it engages with destination stakeholders to address issues (indicator D 1). On the contrary, it is not mentioned the meeting between tourists and local populations that the company could include in the travel package.

The TO responsibility is assessed along the whole tourism value chain, involving all stakeholders: employees, suppliers/partners, customers and society in general. This is not a surprise, since the system is the result of a multi-stakeholders perspective.

The reporting program developed by TOI-GRI follows a certification approach, since it focuses on internal processes and supply chain management. However, unlike other certification programs, in the general Guidelines there are also indicators referring to the product responsibility.

Finally it can be noted that, although the system is well articulated and complete, it is also complex and costly. GRI itself is aware of this issue and acknowledges the difficulty and cost of collecting some data and the unreliability of some

Fig. 5.8 Responsibletravel.com's Logo

information, which lead some enterprises to exclude some indicators in the analysis. However both the General Guidelines and the Sector Supplement seem to be basically designed for big companies, like those that are members of TOI, as Kuoni, TUI, etc., but not for SMEs.

As previously mentioned, the Tour Operators' Sector Supplement has been released in pilot form and at the moment there is no evidence that it is concretely applied by tour operators. On the contrary the General Guidelines are adopted by several companies in different industry sectors.

5.8 Responsibletravel.com

5.8.1 Organisation and Objectives

Responsibletravel.com is the world's first dedicated online travel agent for responsible holidays. Born in 2001, it aims at offering locally distinctive, authentic holidays from across the globe that are better for destinations and local communities. Responsibletravel.com focuses its activity on the distribution of travels that are more responsible and that bring more sources of employment and income for local community, minimising negative impacts on environment and cultural heritage.

Thanks to a platform including over 300 tour operators and 600 accommodation establishments (villas, lodges, B&Bs, hotels, etc.). Responsibletravel.com supports tourists in organising their responsible holidays (Fig. 5.8).

5.8.2 The Reporting System

The system adopted by Responsibletravel.com for selecting and screening its partners and providers is particularly interesting. The travel agency firstly asks its providers to adopt a clear and serious policy that is based on the principles of responsible tourism. Secondly it asks them to demonstrate through appropriate documents that the policy and the business activities carried out meet with the minimum standards defined by the agency (Table 5.13).

In addition, members are required to present a short summary of the responsible travel aspects of every trip and accommodation on the site. They are required to write a paragraph on how the specific holiday makes a difference.

Table 5.13 Responsibletravel.com. The criteria and indicators of the reporting system

1 Policy and Procedure	
1.1	Provide a copy of the company responsible travel policy
1.2	Provide written confirmation that the company policy has been provided to staff, destination suppliers and travellers
1.3	Provide written confirmation that the company has a procedure for responding to travellers complaints & suggestions concerning responsible travel practices
2 Environmental policy	
2.1	Provide evidence of initiatives to reduce waste, and evidence that the company policy requests waste management practice of suppliers (reduce-reuse-recycle)
2.2	Provide evidence that travellers are provided with suggestions for how to reduce water use in their destinations
2.3	Provide evidence that travellers are provided with relevant suggestions to minimise damage to the environment, wildlife and marine ecosystems
2.4	Provide evidence that the company provides suggestions for destination visits to appropriate local projects with direct or indirect environmental benefits
	This is an essential requirement for all trips published on responsibletravel.com
	The company may suggest alternative projects.
3 Social policy	
3.1	Provide evidence that travellers are provided with accurate pre-trip information on the social and political situation in each destination
3.2	Provide evidence that travellers are provided with suggestions of ways to minimise negative impacts on local cultures
3.3	Evidence that the company policy requests destination suppliers to ensure that a local guide accompanies all group visits to local communities
3.4	Provide evidence that the company provides suggestions for destination visits to appropriate local social projects with direct or indirect benefits to the host community. *This is an essential requirement for all trips published on responsibletravel.com*
	The company may suggest alternative environmental projects
4 Economic policy	
4.1	Provide evidence that the company policy requests destination suppliers to employ local people wherever possible
4.2	Provide evidence that the company policy requests destination suppliers, where local standards permit, to make use of local produce, manufacturers and other services
4.3	Provide evidence that travellers are provided with pre-trip suggestions of local services that provide local community benefits (e.g. restaurants, guides, shops, craft markets)

Once accepted, members may use the Responsibletravel.com logo in their marketing and promotion—but only in association with specific trips or accommodation that have met the minimum criteria. For some members, the minimum criteria have been met for all their trips, for others the logo applies only to a select number of them.

Once members have met the minimum criteria, they may continue to add to or improve on these standards as applied to their own products and services. This helps to differentiate individual members and enables them to pursue future change and improvement as appropriate.

It is interesting to note that the audit is made also according to feedbacks sent by travellers, who on their return from their holiday, are invited to comment on their experiences regarding a particular operator or accommodation's responsible tourism policies, in the form of travellers' reviews directly edited in the travel agency website. All of these reviews are sent to the tour companies and hotels as a form of consumer-driven auditing in order to encourage continuous improvement.

The system used by Responsibletravel.com to verify the responsibility of its providers seem to be basically based on the responsible tourism approach, since it aims at ensuring consumers to enjoy a real responsible experience, thanks to tour operators and other providers who are seriously committed towards responsible tourism. This perspective is also confirmed by the attention paid not only on the business policy, but also on the single travels and services.

Although the system covers all aspect of triple bottom line and all main values of responsible tourism, it should include indicators about the relationship between the company and its suppliers, which should be fair, and about the percentage of income that is retained in the destination by the local community.

The system seems to be easily implemented by small and medium enterprises.

5.9 Travelife: Travelife Sustainability System

5.9.1 Organisation and Objectives

Travelife is a training, management and certification initiative based in the Netherlands for tourism companies committed to reach sustainability. Aimed at promoting sustainable practices within the travel and tourism industry, it was launched in 2007 by ABTA—the UK Travel Association and the Dutch ANVR travel association, with the support of Leeds Metropolitan University (UK), Lund University (Sweden) and ECEAT-Projects, a Dutch no-profit organisation promoting sustainability in tourism.

Travelife offers a wide range of services, which include training, management, planning, reporting and B2B tools, believing that a common approach supports a rapid introduction and dissemination of knowledge and desirable practices within the sector. It aims at increasing the quality of travel products, the customer satisfaction and the quality of life in destinations, by bringing together the individual and collective experiences of tour operators, travel agents and their associations and builds upon the experiences of other initiatives within the tourism sector.

In particular, in order to help tour operators and their suppliers, Travelife offers a set of integrated industry-wide instruments and tools, such as:

- System, an international management standard for the implementation of sustainable tourism by tour operators;
- Training, a state-of-the-art course including the best industry practises;

Fig. 5.9 Travelife's Logo

- Planning, setting and monitoring the company commitments through an "action planning system";
- Reporting, an online instrument for tourism organisations to report in- and externally on the accomplishment of specific sustainability actions. It provides an overview of possible actions which correspond with the actions in the "Action Planning" tool;
- Industry wide best practise standards and assessment for tour operator suppliers and destinations

The Travelife partners that initiated the Travelife programme were: ECEAT-Projects (European Centre for Eco and Agro Tourism), Netherlands; ANVR (Dutch Association of Travel Agents an Tour Operators), Netherlands; the Federation of Tour Operators, UK; Leeds Metropolitan University, UK; Lund University, Sweden; the Catalan Ministry of Environment, Spain; the Austrian Ministry of Environment, Austria; the Royal Award Foundation, Denmark, and the Dutch Alps Platform, Netherlands.

Nowadays, Travelife is supported by a growing number of trade associations that have committed their members to implement a sustainability management.

Among its several activities, Travelife has developed the Travelife Sustainability System; a reporting program that has been set up to support an efficient and cost effective introduction of sustainability principles within the tourism industry. The Travelife Sustainability System was developed during the EU supported LIFE project, based on sustainability management concepts developed by UNEP, TOI and earlier experiences of ANVR and ABTA (Fig. 5.9).

5.9.2 *The Reporting System*

The Travelife Sustainability System is a fair and affordable system which helps tourism companies to manage and improve social and environmental impacts by complying with the sustainability criteria. At the end of the process, when a company meets the requirements, it receives a Travelife award.

The program, operated by ECEAT-Projects, is composed of two separate but interrelated versions addressed to travel companies—tour operators and travel agencies—and accommodations. In this book the attention is paid in particular on the system for tour operators.

The system has been designed according to the following principles.

- Transparent, comparable and accountable, for ensuring consistency and comparability of audits and allowing sector wide benchmarking.
- Sector wide: applicable to both small and large companies and covering all types of travel companies including inbound, outbound and retail travel companies.
- International: global availability of online tools in multiple languages.
- Development oriented, which means that step by step companies and their suppliers are supported in implementing gradually the Travelife standard.
- Knowledge based, in order to develop knowledge and skills for the travel industry.
- Supply chain oriented, by taking into account the supply chain of travel companies, including transport, accommodation and excursion providers and offering easy tools to evaluate supplier performances.
- Efficient, by encouraging data sharing among participants, since a common database of suppliers creates efficiencies and avoids duplicating audits.
- Unified communication, by requiring sector-wide logos and messages for communication in travel companies' promotion materials.
- Realistic and practical, in the sense that criteria and actions are based on what travel companies can do.
- Triple bottom line, by integrating of the environmental, social and economic dimensions.
- State of the art, by integrating the major global standards and guidelines, as EMAS III, ISO 14001, ISO 26001, GRI, OECD standard for multinational companies and the Global Sustainable Tourism Standard (GSTC).
- Collaborative, by encouraging involvement of relevant stakeholders and destinations.
- Not for profit, in the sense that Travelife runs on a cost basis and all income will be used to manage and improve the system and the other tools.

According to the principles listed above, the Travelife Sustainability System not only provides criteria and minimum standards for the certification, but also online training and practical tools for sustainability planning, management and reporting. Training and tools should be suitable for companies of any size and cover all management fields of the travel business, including office operations, supply chain, destinations and consumers.

The Travelife Sustainability System is based on a selection of criteria and indicators referring to the whole value chain and grouped into eight areas.

- Sustainable management
- Internal management
- Sustainable Supply Chain Management
- Sustainable Transport
- Sustainable Accommodation
- Sustainable excursions
- Destination
- Customer communication

Table 5.14 gives an overview of criteria and indicators.

Operatively, the system is accessible through a Web based facility (www.travelife.org). Upon compliance with the Travelife standard, the travel company can obtain the "Travelife Partner" or "Travelife Certified" status. In particular, the Travelife procedure implies three steps to be carried out by the company before awarding the certification.

- Stage 1—Travelife Engaged: in the first stage, the company commits itself and introduces the appropriate management procedures. An appointed Sustainability Co-ordinator obtains the personal Travelife certificate and elementary steps and good practices are implemented. During this "engagement" stage the company itself is not rewarded yet.
- Stage 2—Travelife Partner: based on a sustainability report and in compliance with basic requirements assessed by Travelife, the company reaches the Travelife Partner status. This status enables the company to communicate its sustainability commitment and achievements with the support of the Travelife Partner logo. The company is now qualified to work step-by-step towards the final stage.
- Stage 3—Travelife Certified: in the third stage, the company's compliance with the international Travelife standard for travel companies is evaluated by an independent auditor based on an on-site assessment. The requirements are stronger than the basic requirements of the Travelife Partner stage. After successfully completing the third stage, the company is be awarded with the Travelife Certified status.

The system developed by Travelife, as the Tour Operators' Sector Supplements of TOI-GRI and the CSR Reporting Standards of KATE-TourCert, is mainly based on the certification approach, since it assesses the responsibility in the internal processes and activities of the company and in the supply chain. However the Travelife Sustainability System does not forget to consider initiatives aimed at increasing travel awareness about responsible behaviour to be taken at the destination.

The social and environmental dimensions are well developed, while the economic aspect would need further improvement, the only indicators that can be associated to the economic dimension is the preference for local providers in order to support local economy.

The Travelife Sustainability System is presently applied by more than 450 tour operators, members of international tour operator associations such as the Dutch ANVR, the British ABTA/FTO, the German AS, the Scandinavian SRF (Sweden), the German DRF (Denmark), the Finnish SMAL/AFTA. Among these tour operators, there are important companies such as TUI, Kuoni and Thomas Cook. Therefore the high number of applicants seems to confirm the success of the system, that, together with the program of KATE-TourCert, is the most widespread in the European Travel Industry. However, the high number of indicators induces some doubts about the applicability of the system by SMEs.

Table 5.14 Travelife. Criteria and indicators of the reporting system

1	**Sustainable management**
1.1	**Engagement Solution**
1.1.1	A sustainability coordinator is appointed and his/her role and activities are defined
1.1.2	Sustainability-related tasks and responsibilities are part of the company procedures and processes
1.1.3	Appropriate responsibilities are assigned to the company's staff for actions related to the sustainability policy
1.1.4	A sustainability mission statement is defined
1.1.5	The sustainability coordinator has successfully passed the Travelife training
1.1.6	Management and other staff members have successfully passed the Travelife training
1.1.7	The Sustainability coordinator, staff members and management have received additional education related to sustainable tourism management (e.g. ISO 14001, University courses, training for guides and others)
1.1.8	Staff members are regularly updated on the company's sustainability progress (e.g. newsletters, intranet and other internal communication channels used by the company)
1.1.9	The organisation is actively involved in external forums and working groups which are supportive to sustainable tourism
1.1.10	All experiences and best practices are concerning sustainable tourism are shared with other organisations and/or on the Travelife website
1.1.11	The organisation makes donations to (local) institutes, nature protection organisations or other sustainable local/national initiatives
1.2	**Baseline Assessment**
1.2.1	The company has identified the nature and the number of suppliers whom is working with (directly or indirectly) and their sustainable performance (e.g. through the Travelife baseline assessment)
1.3	**Policy**
1.3.1	The organisation has a formal, written policy statement on sustainable tourism
1.4	**Action Plan**
1.4.1	The organisation has a sustainability action plan
1.5	**Monitoring and Evaluating**
1.5.1	The organisation developed an (internal) management system to monitor and evaluate the progress of the sustainable management system (e.g. by extension of existing management programmes with sustainability information and files)
1.6	**Reporting**
1.6.1	The Travelife sustainability report is submitted every 2 year
1.6.2	A public sustainability report is published annually
2	**Internal Management**
2.1	**Social Policies**
2.1.1	Trade union membership, collective negotiation agreements and representation of members by trade unions is allowed
2.1.2	All employees of the organisation have a written employment contract
2.1.3.	The organisation contributes to a health-care insurance for all employees

(continued)

Table 5.14 (continued)

2.1.4	The organisation offers schemes for pregnancy—maternity and parental leave for all employees.
2.1.5	The organisation contributes to a disability-risk insurance of all employees
2.1.6	The organisation has a pension scheme for all staff members
2.1.7	The organisation contributes financially to education and training of its employees
2.1.8	The organisation has other arrangements for its employees
2.1.9	The risk of corporate accidents and illness is minimised
2.1.10	Physical and mental wellbeing of staff in the main office and in destinations is arranged in compliance with national legislation
2.1.11	Staff changeover does not exceed the national average
2.1.12	The organisation gives preference to the assignment of local staff in destinations, incl. management positions
2.1.13	The organisation ensures that no discrimination based on ethnic background, race, religion or sex will occur in staff employment, training/education and/or access to superior employment levels
2.1.14	There is an effective complaints procedure and possibility to claim for compensation for all employees
2.1.15	Training aimed to maintain staff competencies are stimulated at all levels of responsibility
2.1.16	The organisation offers the possibility to present important labour matters to superior management
2.1.17	Disciplinary procedures are documented and all staff is informed on them
2.2	**Internal Environmental Management**
2.2.1	The printing company works with a certified environmental management system (>50 % of total corporate printing in kilo's or pages)
2.2.2	Brochures are printed on environmentally friendly paper for more than 50 % of total corporate use (in kilo's or pages)
2.2.3	Efficient use of brochures is stimulated. For example with use of a demand system, barcodes to verify stocks or container deposits
2.2.4	Purchasing policy gives preference to sustainable products and services
2.2.5	More than 50 % of paper use in the office consists of sustainable paper
2.2.6	At least 50 % of coffee and team provided in the office consists of fair trade or organic produce, or has sustainable certification
2.2.7	Other office supplies (maps, pens, furniture, etc.) are produced sustainably
2.2.8	Products are purchased in bulk in order to reduce amount of materials
2.2.9	Use of biodegradable cleaning materials and environmentally certified cleaning companies
2.2.10	Catering works with sustainable purchasing, transport, waste etc. and offers a maximum possible of organic food
2.2.11	Presence of other positive examples of sustainable purchasing
2.2.12	The organisation is actively engaged in waste reduction
2.2.13	The organisation has implemented an effective policy for separation and reuse of waste
2.2.14	Glass is recycled
2.2.15	Paper/cardboard is recycled
2.2.16	Plastic is recycled
2.2.17	Toner/ink is recycled
2.2.18	Organic waste is recycled
2.2.19	Batteries are recycled

(continued)

Table 5.14 (continued)

2.2.20	Other waste is recycled
2.2.21	The organisation complies to the national legislation concerning waste disposal
2.2.22	The organisation has an energy-saving policy
2.2.23	At least 75 % of energy efficient lighting is installed in the organisation
2.2.24	An automatic switch on/off system with for example timers or movement sensors is operational
2.2.25	Is all equipment switched off after office hours (not on "standby")
2.2.26	Does the organisation use low energy equipment?
2.2.27	Other measures, not indicated above, are taken
2.2.28	At least 50 % of total energy consumption is sustainable energy
2.2.29	CO_2 emission from the use of fossil energy for the offices is compensated
2.2.30	The organisation has an active policy to reduce water consumption
2.2.31	Flow restrictors are installed in min. 75 % of the taps
2.2.32	Aerators are installed
2.2.33	Percussion taps are installed in min. 75 % of all taps
2.2.34	Push buttons or other water saving equipment is installed in the toilets
2.2.35	Other water-saving examples are available
2.2.36	Grey water is recycled and treated before use
2.2.37	The wastewater is treated the correct way, according to national legislation
2.2.38	Staff is reminded regularly to save water
2.2.39	The organisation has a mobility policy for the staff
2.2.40	Business flights of staff are compensated
2.2.41	Employees are stimulated to use public transport or sustainable transport means like bicycles
2.2.42	Corporate policy aims to reduce the transport necessary by tele-work, tele/video meetings and other means
2.2.43	Compensation for private or lease cars is limited to clean energy cars category A or B
2.2.44	All employees are invited to participate in training or receive information (posters/email) on social and environmental policies
3	**Sustainable supply chain management**
3.1	**Selection of new suppliers**
3.1.1	New suppliers are also selected on their sustainability
3.1.2	Suppliers that work with children younger than 14 years are not selected
3.2	**Incentives for current suppliers**
3.2.1	Current suppliers should meet with the basic criteria
3.2.2	Suppliers are stimulated to implement the Travelife
3.2.3	The Best Practice standards are distributed to all suppliers
3.2.4	Suppliers are encouraged to monitor and evaluate their organisation on a regular basis with the Travelife
3.2.5	Manuals and other information for more sustainable management are distributed to suppliers on a regular basis
3.2.6	Suppliers are advised on specific training tools
3.2.7	Suppliers have participated in local or (inter)national sustainability training and education
4	**Sustainable Transport**
4.1	**Coach**
4.1.1	Collaboration with certified coach transport companies are developed

(continued)

Table 5.14 (continued)

4.1.2	Transport providers are encouraged to implement a sustainable driving style by means of communication of the "eco-driving" technique or international equivalents
4.2	**Less polluting transport to destinations**
4.2.1	Sustainable (public) transport is offered to the point of departure for the international journey
4.2.2	(part of) The transport offered to the destination is sustainable
4.2.3	CO_2 compensation for the international transport is included in the package price
4.2.4	The organisation does not offer flights: For distances <700 km; if total holiday package is less than 8 days, no flights or distances 700><2,000 km are offered; if total holiday package is less than 14 days no flights for distances >2,000 km are offered
4.3	**Transport in destination**
4.3.1	Suppliers and local agents are stimulated to offer sustainable transport for local transfers and excursions
5	**Sustainable Accomodation**
5.1	**Accomodation**
5.1.1	Preference is given to accommodations that function sustainably
5.1.2	Preference is given to accommodations with internationally acknowledged and/or Travelife certification
5.1.3	The tour operator gives preference to accommodations which are operated by/owned by local entrepreneurs
5.1.4	In countries/regions renowned for sexual exploitation of children the accommodation suppliers are informed on how to recognise the signals of sexual exploitation and other related criminal activities, and on possible means to prevent it
5.1.5	In countries/regions renowned for sexual exploitation of children, the accommodation contracts include clauses which enable the tour operator to end the contractual agreement prematurely if the accommodation supplier does not take adequate measures to prevent sexual exploitation of children
5.1.6	Personnel in countries renowned for sexual exploitation of children receive relevant training
5.1.7	Accommodations are stimulated to offer locally produced and sustainable souvenirs
5.1.8	Local suppliers do not promote souvenirs which are produced with materials from threatened flora and fauna species as indicated in the Cites treaty and the IUCN "Red List"
5.1.9	The tour operator collaborates with hotel managers, sales points within the accommodations and shops within the destination to assure that the local offer of souvenirs is sustainable. Moreover, clients are informed on the reasons
6	**Sustainable excursion**
6.1	**Excursion**
6.1.1	The development/production of excursions offered to tourists cause minimal or no damage to the natural environment, infrastructure, etc.
6.1.2	The organisation offers no products or services that are damaging to humans, animals, plants, natural sources (e.g. water/energy), or which are socially/culturally unacceptable
6.1.3	Activities, products, product parts that are not according to national/international legislation and/or agreements are not offered by the organisation
6.1.4	No promotion is made for hunting, unless it takes place with consideration of all environmental interests and with government permission

(continued)

Table 5.14 (continued)

6.1.5	Excursions and activities which are supportive to local communities are developed and offered to consumers
6.1.6	Excursions and entertainment with attention for child prostitution or aimed at child prostitution will not be promoted and will not be offered
7	**Destination**
7.1	**Destination**
7.1.1	In product development preference is given to sustainable destinations (e.g. no traffic, sustainable city planning and infrastructure)
7.1.2	When composing the tours the organisation gives preference to destinations which are easily accessible by sustainable means of transport
7.1.3	Extremely vulnerable destinations (e.g. Antarctica, Galapagos) will not be visited unless the visit is ecologically responsible
7.1.4	The organisation supports initiatives that improve the relationships between accommodations and local producers, among which the producers of local food products and souvenirs etc. (e.g. initiatives to improve quality level, logistics, transport
7.1.5	The tour operator collaborates with other tour operators, nature protection organisations, government organisations in order to prevent trade in threatened flora/fauna and archaeological/cultural heritage
7.1.6	Excursions that promote local crafts and local (food) production methods are developed and promoted
7.1.7	The tour operator influences local government (when possible together with other tour operators and stakeholders) concerning sustainability, destination management and planning, use of natural sources and socio-cultural issues
7.1.8	Local initiatives to support sustainability are included in excursions (e.g. regarding biodiversity, social projects and nature conservation
7.1.9	The tour operator supports the development of adequate waste (water) management, infrastructure and facilities in beach destinations
8	**Customer Satisfaction**
8.1	**Customer Satisfaction**
8.1.1	The corporate sustainability policy is accessible to consumers and clearly communicated
8.1.2	The corporate results regarding the sustainability policy are communicated to consumers
8.1.3	Consumers are provided with information on sustainable tourism and sustainable behaviour (e.g. respectful attitude regarding local socio-cultural and natural environment). Recommendations on how to contribute to this
8.1.4	Consumer is informed on the natural and socio-cultural environment in the holiday destination
8.1.5	Clients receive guidelines and/or codes of conduct for attractions with animals (in captivity)
8.1.6	Clients receive information on sustainable behaviour on the beaches of the holiday destination. Clients receive guidelines/code of conduct for all marine activities, e.g. observing whales, dolphins, turtles, etc.
8.1.7	Clients receive guidelines/code of conduct for sustainable snorkeling and diving
8.1.8	Clients receive guidelines/codes of conduct for sustainable sports fishing
8.1.9	Clients receive guidelines/code of conduct for suitable behaviour in cultural destinations/sites
8.1.10	Clients receive guidelines/code of conduct on wildlife tourism
8.1.11	Clients receive guidelines/code of conduct for encounters with native and traditional cultures

(continued)

Table 5.14 (continued)

8.1.12	Clients receive guidelines and/or code of conduct for sustainable winter sports activities
8.1.13	Clients receive general guidelines for sustainable behaviour and/or code of conduct for local activities
8.1.14	Clients are informed on several transport means to reach the destination, with the environmental impact and (more) sustainable alternatives
8.1.15	Clients are informed on the possibilities for CO_2 reduction/compensation, in particular for packages including transport by airplane
8.1.16	Sustainable accommodations and excursions are promoted with logo's or other promotional materials, clearly visible and recognisable to consumers
8.1.17	Clients are informed on all relevant legislation concerning the purchasing, sales, in- and export of historic or religious artefacts and articles containing materials of threatened flora and/or fauna
8.1.18	Clients are stimulated to use meals also in local restaurants, not solely in the holiday accommodation. -
8.1.19	Local staff and local representatives promote sustainable local excursions and local produce
8.1.20	Clients are encouraged to donate to local charity and sustainable initiatives (e.g. nature conservation and social projects)
8.1.21	Client satisfaction is structurally measured
8.1.22	Sustainability is an integral part of the research into client satisfaction (incl. complaints)

5.10 Comparison of Reporting Systems: Commonalities and Differences

The reporting systems described in the previous sections can be first compared according to the type of approach they follow: the certification approach versus the responsible tourism approach.

Figure 5.10 shows that while the CSR Reporting Standards of KATE-TourCert and the Tour Operators' Sector Supplement of TOI-GRI are completely based on the certification approach, the only program following a totally responsible tourism perspective is the Responsible Tourism Standards of AITR, which assesses the package tour. The other methodologies are the result of the integration of both approaches, and at a different level they try to combine a market-driven perspective paying greater attention to internal processes, activities and supply chain developed by the tour operator. The systems of ATES, ATT and Responsibletravel.com are more linked to the program of AITR, since they share the objective to identify and recognise those tour operators specialised in responsible tourism travels. On the contrary, the schemes developed by QUIDAMTUR and Travelife are closer to KATE and TOI-GRI, since they put more emphasis on the responsibility along the internal processes and the supply chain, although they do not forget to ensure consumers and raise their awareness.

The difference of perspective that characterises the systems considered in the analysis is also confirmed in Tables 5.15 and 5.16, which facilitate the comparison in terms of objectives, assessment procedure, indicators and type of auditing. A first

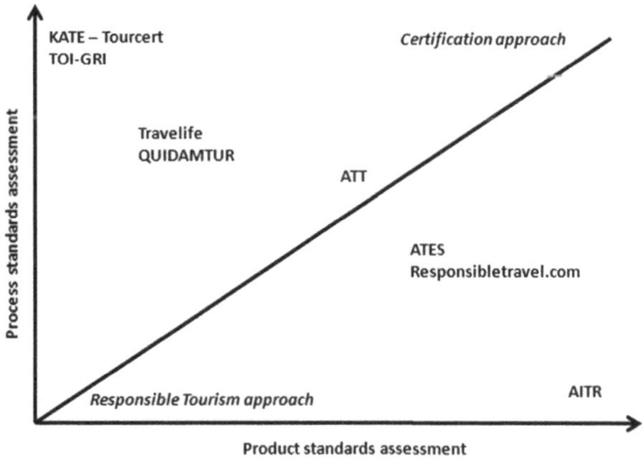

Fig. 5.10 Classification of reporting systems according to the certification approach and the responsible tourism approach (Authors' own creation)

element to highlight is that, unlike other programs—which generally concentrate on the company through a performance or process-based approach—the Responsible Tourism Standards of AITR is the unique system that focuses on the travel package.

Another interesting aspect explained (Tables 5.15 and 5.16) is the type of auditing. KATE, ATT and Travelife are the only organisations that, after a self-evaluation procedure, requires an audit by an external, independent third party that are in charge of granting the certification (for example, TourCert for the CSR Reporting Standards of KATE, AFNOR for the ATR System of ATT and another independent agency for Travelife). This ensures a greater objectivity, transparency and credibility also in the eyes of consumers and public opinion in general.

The other programs only require a self-evaluation by the company, followed by a ratification by the association and eventually by reviews of consumers.

Another significant element is represented by the level of attention paid by programs to the triple bottom line, i.e. to the environmental, social and economic dimensions.

As highlighted in Table 5.17, AITR and ATES are the organisations that, more than their colleagues, give extremely importance to the social aspect, in particular to the host community involvement and the encounter between tourists and local people. They have developed a set of specific indicators to assess the relationship between the company and the local community, which should be based on values of fairness, respect and equity, and the cultural exchange between tourists and local people, that the tour operator should include in the travel. This focus depends on the fact that the members of these associations are specialised in responsible tourism, for which these aspects are strategic.

The socio-cultural dimension is also developed by ATT, through a set of indicators proving that the tour operator respects labour rights and gives tourists detailed information about the socio-cultural background in the destination and the Ethic Chart of the association. TOI-GRI and Travelife, on the other hand, assess in particular the tour operator's commitment to cooperate with the community and find a solution for the issues of the destinations, to appropriately inform consumers and avoid discriminations and employees' exploitation.

Table 5.15 General comparison of reporting systems (follows)

	AITR	ATES	QUIDAMTUR	KATE
Objectives	– Distinguish r.t. from other forms of tourism – Protect consumers, ensuring that travels respect principles of r.t. – Protect AITR members from other TOs using r.t.. only for marketing reasons – Provide AITR members with a set of minimal operational standards	– Share a common vision among members – Promote transparency, informing travelers that ATES members respect r.t. – Enhance the credibility of ATES members with public opinion	– Asses the level of responsibility of TOs and of tourism enterprises along the supply chain – Develop an unambiguous tool, universally applicable, to evaluate the sustainability of tourist services developed	– Help TOs to build an integrated responsible strategy – Give transparency on ecological, social and economic impacts of business – Increase credibility among customers
Evaluation object	The travel package	The TO: travel organisation and project carried out	Structure and internal process of every company of the supply chain	The TO: its internal and external CSR
Criteria	13	10	54	8
Indicators	23	29	98	57
Auditing	Self-evaluation and approval by AITR	Self-evaluation and approval by ATES	Self-evaluation, assessment by clients, assessment by local experts	Third-party audit
Current situation	Not applied by any tour operator	Applied by almost 23 tour operators, members of the association	Not applied by any tour operator	Applied by almost 70 tour operators

TOs tour operators
r.t. responsible tourism

KATE, TOI-GRI, Travelife and partially QUIDAMTUR are instead those systems that have better developed the environmental dimension, through minimum standards that ask the tour operator to respect and preserve the environment both along the tourism value chain, at the destination and within the company (internal processes). However, also ATT and AITR do not neglect this aspect: the French association requires its members to adopt an environmental policy and to raise awareness of suppliers and consumers, while the Italian organisation encourages its associates to avoid fragile areas, to prefer public transports in the destination, etc.

Table 5.16 General comparison of reporting systems

	ATT	TOI-GRI	Responsibletr.	Travelife
Objectives	– Support the development of sustainable and quality tourism – Monitor the economic impacts in destination – Respect heritage, environment and culture – Develop public awareness transparently	– Create a reliable reporting system, which can be used by enterprises of all sizes and nationality	– Distribute responsible travels, which create new income and job opportunities for local communities and minimise negative impacts – Select and control partners	– Help TOs and their suppliers to operate according to the principles of r.t. – Inform customers – Increase product quality, customer satisfaction and well-being in destinations
Evaluation object	The TO internal processes and travel organisation	The TO: its internal and external CSR	The TO: policy, activities and travel characteristics	The TO: environmental and social performance
Criteria	16	5	4	17
Indicators	24	126	14	139
Auditing	Third-party audit	Self-evaluation	Self-evaluation, consumers' reviews and approval by the agency	Third-party audit
Current situation	Applied by almost 15 tour operators, members of the association	Not applied by any tour operator	Applied by partners and providers of the travel agency	Applied by several tour operators (450 tour operators have introduced the system)

TOs tour operators
r.t. responsible tourism

While, despite some differences, the social and environmental dimensions are considered by all programs, the economic aspect is particularly developed by TOI-GRI, QUIDAMTUR and KATE. They require tour operators to demonstrate the impacts of their activity on the local economy and to describe their financial background. Although the last details may seem to be not so significant in assessing the business responsibility, they help to identify the size of the enterprise and to

Table 5.17 Comparisons of reporting systems in relation to the triple bottom line

	Socio-cultural	Environmental	Economic
AITR	5	3	1
ATES	5	2	1
QUIDAMTUR	3	4	4
KATE	2	5	3
ATT	4	3	1
GRI E TOI	4	5	5
Responsibletravel.com	3	2	1
Travelife	3	5	1

The measurement scale ranges from 0 to 5

Table 5.18 Comparisons of reporting systems in relation to the tourism business stakeholders

	Employee	Clients	Providers	Host community
AITR	0	5	0	5
ATES	0	5	0	5
QUIDAMTUR	3	4	5	3
KATE	5	4	5	2
ATT	2	5	3	2
GRI E TOI	5	4	5	3
Responsibletravel.com	0	4	0	2
Travelife	4	5	5	3

The measurement scale ranges from 0 to 5

explain some potential differences among companies in terms of environmental, social and economic performance and amount of investments.

However, considering all reporting programs, the system of TOI-GRI seems to develop, better than the other schemes, all three dimensions of responsibility with a sufficient set of indicators.

Since the responsibility of a company should be assessed in relation to all main groups of stakeholders, it is also useful to distinguish reporting systems according to the involvement of employees, providers, distributors, consumers, local community, etc. in the activities carried out by the business.

As shown in Table 5.18, all programs pay particular attention to consumers, asking tour operators to provide them with comprehensive information about the destination, to raise their awareness about responsible behaviour and to ensure transparency and safety. While great importance to the local community is given in particular by the systems of AITR, ATES and partially ATT, a significant role to employees and providers is above all recognised by KATE, TOI-GRI, Travelife and QUIDAMTUR. This is not so surprising, since the different perspective these systems follows: the first two schemes focus on responsible tourism, for which the host community should play a key role, while the other alternatives are based on the assessment of internal processes and relationships within the supply chain.

It should be noted that, regardless of the approach, a comprehensive reporting system should consider the business responsibility in relation to all main groups of stakeholders.

The different importance given to stakeholders explains the divergence in terms of technical documents required to tour operators by the systems. While the use of a questionnaire for analysing the customer satisfaction is a requirement included in all systems, the evaluation sheet for suppliers and the questionnaire for the employee satisfaction are a condition put by some schemes only (for example, those of KATE, TOI-GRI and QUIDAMTUR).

The comparison that has been carried out according to several aspects, has underlined that, although reporting systems share some common elements, they are heterogeneous. Not only they differ according to the assessment procedure, the number and typology of indicators, but also in terms of objectives and approach. This last element is particular decisive, since it influences all other characteristics. On one hand, the programs of AITR, ATES, ATT and Responsibletravel.com, that are more focused on ensuring and promoting responsible tourism; on the other, the schemes of KATE, TOI-GRI, QUIDAMTUR and Travelife, which are more interested in pushing corporate social responsibility integrated by the company in its websites.

It is difficult to say which system is the best one, since each program has its strengths and weaknesses according to what can be defined as an "ideal system", i.e. a system that takes into account all aspects of responsibility in tourism (social, cultural, economic and environmental), that assesses it within the enterprise as well as along the whole supply chain and the single travel package and that involves all stakeholders (internal staff, suppliers, clients, local population).

At this point, it would be important to discuss whether the poor diffusion of some systems and, on the contrary, the popularity of others, can represent an evidence of a better quality of some initiatives in comparison to others.

Systems as those of Travelife and KATE-TourCert seem to have found a large consensus in the tourism industry, since they are applied by a high number of European companies. In addition, they have obtained the support of other organisations and started to co-operate with other reporting systems. For example, both Travelife and TourCert have recently subscribed a mutual recognition agreement with Fair Trade Tourism (FTT), the southern Africa's leading responsible tourism non-profit organisation that operates a certification scheme supporting tourism enterprises to optimise their environmental, social and economic impact. On the one hand, the agreement will allow any tour operator certified by TourCert or Travelife to receive the full number of possible points when using an FTT-certified tourism business as a supplier, thus by-passing the condition required by TourCert and Travelife to conduct an individual check of this supplier. On the other, FTT will automatically recognise tour operators and travel agencies certified by TourCert and Travelife as being "pre-approved" to sell Fair Trade Holidays, a new concept developed for the international trade in tourism services, subject to related marketing efforts and sales transactions being periodically audited by an independent certification body appointed by FTT.

Therefore it should be interesting to analyse if the high number of applicants and the support and the cooperation granted by other organisations to Travelife and KATE-TourCert depend on the quality of their systems or on other variables, such as partnership development, effective communication and promotional tools that make these systems more acknowledged than competitors. It is likely that this success is a combination of all these aspects.

5.11 Issues and Open Questions

A number of critical issues emerge from the analysis.

- The first critical aspect concerns the object of the assessment, i.e. the enterprise vs. the travel package. Apart from the AITR system, all methodologies analysed basically focus their attention on the enterprise. This decision is also linked to the type of approach adopted. Whether certifying the enterprise can be more useful for the tour operator itself, its suppliers and shareholders, the focus on the travel package can be more effective in showing consumers the engagement of the tour operator and in making them aware of responsible tourism practices. This is the reason why many methodologies try to combine both approaches, following a path that may represent the best alternative. The integration of both perspectives enables companies to pay attention and monitor both the responsibility along the internal process and the supply chain (the CSR) and the product adherence to values of responsible tourism. In addition, it should be noted that the flop of the system completely based on the responsible tourism approach, i.e. the Responsible Tourism Standards of AITR, which at the moment is not applied by any tour operators, seems to demonstrate the failure of the market-driven perspective, in favour of the certification approach or of the integration of both perspectives.
- The second important issue is related to the size of the tour operators involved in the assessment. In the light of the difficulties that small- and medium-sized businesses face in investing in CSR and in adopting a reporting system, methodologies like the Tour Operators' Sector Supplement of TOI-GRI, the CSR Reporting Standards of KATE-TourCert or the Travelife system do not seem to be applied easily by small tour operators. They are particularly complex and costly in terms of time and human and financial resources needed to join the assessment and collect the information. The issue should not be undervalued, since most companies of the tourism industry may not be able to assess their business responsibility. Therefore, the solution could be represented by a system that is easier and more compliant and based on less rigorous minimum standards. However, while flexibility in the assessment procedure is desirable for SMEs, an oversimplification could undermine the role of reporting programs. Consequently, it is important to find a balance between the needs of small businesses and the effectiveness of reporting systems.

- The third aspect concerns the tour operator specialisation in responsible tourism: the involvement of individual tour operators vary considerably and it is influenced by the type of holiday packages offered (and hence the type of consumers), by the destinations served and by their brand identity (in terms of reliability). From this point of view, some of the systems address to tour operators specialised in responsible tourism only (e.g. AITR, ATES, Responsibletravel, ATT-ATR). Other programs (KATE, TOI-GRI, Travelife, QUIDAMTUR) are open to all tour operators, since their purpose is to assess the corporate social responsibility of any enterprise. However the decision to focus only on responsible tourism operators implies a preliminary selection of companies who can have access to the scheme and therefore the definition of the criteria for establishing the level of specialisation in responsible tourism. This can be expressed in different ways: for example, through the percentage of responsible travel packages on total packages sold or the percentage of turnover deriving from responsible packages on total turnover. In any case, the decision may either include tour operators that are not truly focused on responsible tourism or limit too much the access to the system, allowing few companies to participate. Then this phenomenon may limit tourists, and in particular those interested in responsible tourism, in clearly recognising tour operators specialised in this form of tourism.
- The last critical aspect concerns the assessment procedure (self-evaluation vs. second-party or external audit). As already discussed, an external audit assures a neutral, independent and transparent evaluation according to established standards. The self-evaluation, on the contrary, does not ensure that the statements made by the enterprise are objective and true. Generally speaking, whether the external auditing is the best solution, it is often the cost of the process that forces organisations to choose the other two options.

Together with these critical aspects, there are other two issues that have to be taken into account. The first one concerns the difficulties for some tour operators to work with providers that can be defined as "responsible". In developing countries, services provided by local suppliers cannot often accomplish the CSR standards, because of the lack of knowledge and adequate tools. Some practices that are now common in developed countries—waste sorting and recycling, respect for child labour, use of alternative energy source, etc.—are not applied in developing countries. Unfortunately, this situation does not allow tour operators to respect the minimum requirements for an appropriate supply chain management. The obstacle can be overcome both by adopting a tolerance threshold in standards' application and by encouraging tour operators to support local suppliers with knowledge transfer and training.

The second issue regards the difficulty to translate some principles in measurable indicators. For example, although it is important to assess the tour operator commitment to involve the local community in travel planning and to promote the encounter between tourists and residents, it is not an easy task to build a set of appropriate indicators to check these aspects. For example, it should be necessary to clarify what the system means for "encounter between tourists and local communities"; how many meetings are required for ensuring a real understanding of the destination; which local people are so representative of the culture and issues in the community to justify an interesting and profitable meeting that really makes travellers aware of the specificity of the destination.

Chapter 6
Business Case-studies

Abstract This chapter proposes a set of business case-studies showing how reporting systems can be successfully adopted and awarded by tour operators. The experiences of three different tour operators have been summarised (the German *a&e erlebnis:reisen*, the French *Atalante* and the Dutch *Natuurreizen*), who have implemented one of the programs described in Chap. 5: the Reporting Standards of KATE-TourCert; the ATR system of ATT and the Travelife Sustainability System, respectively. Since these tour operators differ in terms of size, supply specialisation, target market, etc., the case studies allow readers to see how every company has developed its responsible path, the policies and actions they have implemented (in favour of the environment, the host community, the local economic growth, the employees) and their projects for the future.

6.1 Introduction

In this chapter three significant business case-studies are proposed in order to show how some tour operators have been able to successfully adopt a responsible path into their core-business, internal processes and activities and to implement a reporting system for assessing their commitment and performance.

The case studies represent companies who act in a responsible way—following responsible tourism or adopting CSR—without compromising their economic-financial performance and making their commitment a source for building consumer trust and new competitive advantages.

In particular, three European tour operators have been chosen (the German *a&e erlebnis:reisen*, the French *Atalante* and the Dutch *Natuurreizen*), who have adopted one of the systems considered in the analysis: the CSR Reporting Standards of KATE-TourCert, the ATR system of ATT and the Travelife Sustainability System, respectively.

M. Manente et al., *Responsible Tourism and CSR*, CSR, Sustainability, Ethics & Governance, DOI 10.1007/978-3-319-06308-9_6,

Tables 6.1, 6.2 and 6.3 show the main characteristics of these companies, in particular in terms of processes they have adopted to integrate responsibility into their core business and to implement the reporting program.

These tour operators differ in terms of size, target market, product supply and responsible process carried out, highlighting the great variety that characterises tourism companies engaged in promoting sustainable tourism development. In addition, since tour operators are both small and medium-sized enterprises (from a minimum of 7 to a maximum of 70 employees), the analysis enables readers to check how operators of different size approach CSR and responsible tourism. For example, the experience of the micro tour operator a&e reiseteam seems to demonstrate that also very small businesses can successfully implement a reporting program.

6.2 a&e erlebnis:reisen

a&e erlebnis:reisen is a German tour operator based in Hamburg, born in 1991 and specialised in responsible tourism.

In 2009 the company decided to formally adopt CSR and to implement a process for the evaluation and certification of its responsibility.

6.3 Atalante

Atalante is a French tour operator based in Lyon, founded in 1986 and specialised in slow tourism, in particular in walking holidays. It has three offices open to the public: two in France (Lyon and Paris) and another one in Belgium (Bruxelles).

Atalante is a cofounder of ATT, the French Association of Thematic Tour Operator that is born in 2004, and it has awarded the ATR certification in 2008.

6.4 SNP Natuurreizen

SNP Natuurreizen is a Dutch tour operator born in 1984 and based in Nijmegen.

It offers nature-oriented, active and adventure tours that, since the company's opening, are based on sustainability values. The SNP was founded by biologists.

SNP has integrated CSR in its business and in 2012 it implemented and awarded the Travelife Sustainability System for the first time.

Table 6.1 a&e erlebnis:reisen's sheet

Logo	
N° of employee and average annual turnover	Employees: 7 Turnover: 1,843,000 €
Main destinations offered by the tour operator	80 destinations across the world Main destinations are Indochina, India and Central Asia and Latin America Other destinations proposed with a lower frequency are Africa, USA, Europe and Antarctic
Average n° of annual clients	611 for 11.584 total nights
Some information about the demand	All clients are interested in experiencing a responsible travel that is different from traditional holidays offered by generalist tour operators. Most clients (432) travel in groups of minimum 6—maximum 15 participants. The other clients (179) are free individual tourists that travel with their family or with friends and that want an ad-hoc holiday to be independently experienced
Motivation for the decision to implement CSR or to specialise in responsible tourism	The decision to focus on responsible tourism and to implement CSR is based on the specific mission of the tour operator: a&e aims at organising responsible travel packages that do not damage the environmental resources and the cultural heritage and that encourage a positive meeting and exchange between tourists and local people, together with a deeper understanding of issues affecting destinations especially in developing countries Holidays proposed by a&e want to be not only an experience of fun and relax but also a human and ecological experience
Policies and actions for the environmental preservation	a&e: – Buys energy from a provider who certifies the CO_2 emission minimisation and who respects environmental standards – Asks employees to use the train (and not the airplane) for business trips, if possible – Asks employees to use public transports for going to work – Buys coffee, tea, milk, drinks from producers of organic agriculture

(continued)

Table 6.1 (continued)

Logo	a&e erlebnis : reisen
	– Prints catalogues and brochures on FSC certified paper (Forest Stewardship Council) – Uses recycled paper (90 % of paper is recycled) – Sends to clients documents through e-mail, if possible – Compensates CO_2 emissions – Avoids to use airplane for short trips, opting for other means of transports – Uses airplane only for long haul trips of more than 14 days – Encourages clients to use public transports for reaching the departure airport (or the station) – Uses public transports for transfers during the holiday in the destinations – Selects local providers according to their environmental behaviour that is appropriately checked – Financially supports environmental projects concerning the destinations that are offered (for example WWF projects)
Policies and actions in favour of the local community in destinations	Travel packages are organised in a way that they are positive not only for tourists but also for the community. In particular: – All travel packages are conceived and organised in cooperation with local partners – Travel plans include experience for understanding local culture and traditions – a&e financially supports social projects aimed for example at improving education and training in the destination
Policies and actions towards employee s and clients	A&e recognises for its employees: – Flexible working hours schemes in order to meet personal needs – Free access to coffee, tea, drinks, etc. – Training courses For clients, a&e: – Gives detailed information before the departure about the destination and the responsible behaviour

(continued)

Table 6.1 (continued)

Logo	
	– Supports clients in choosing the best holiday solutions – Monitors customer satisfaction, through an ad-hoc questionnaire. 84 % of clients are satisfied
Policies and actions for supporting the economic growth in destinations	a&e: – Works only with local providers (accommodations, tour guides, etc) in order to contribute to the local economy; 36 % of the holiday price is addressed to pay local providers and partners – Financially supports local development projects
Reporting system	Since 2009 a&e is "CSR Tourism" certified (CSR Reporting Standards of KATE-TourCert). The certification must be updated every 2 years
Projects and investments for the future	– Improve the environmental performance though energy saving (for example by inviting employees to adopt eco-friendly behaviour in their office), waste sorting, paper reduction, catalogue page reduction – Improve training courses for employees – Other measures

Source: authors' reworking on the basis of a&e erlebnis:reisen CSR Report—Nachhaltigkeitsbericht 2012, retrieved from http://www.ae-erlebnisreisen.de/nachhaltigkeit

6.5 Some Final Considerations

The case-studies analysed highlight that different motivations and approaches can be followed by companies that are committed to be more responsible from an economic, social and environmental point of view.

While for Atalante and Natuurreizen the commitment towards responsible activities answer to the specific wishes and attitudes of the business founders, for a&e erlebnis:reisen the engagement is the unique path to be followed, given its specialisation in responsible tourism.

However, in all cases the commitment seems to be directly or indirectly moved by the operator's awareness towards ways of travelling that are more responsible and respectful of people and places. Therefore operator's awareness, sensitiveness

Table 6.2 Atalante's sheet

Logo	Atalante
No. of employee and average annual turnover	Employees: 23 Turnover: 10,500,000 €
Main destinations offered by the tour operator	60 destinations and 300 itineraries across the world: Africa, North and South America, Asia and Europe
Average no. of annual clients	Clients: 6,000
Some information about the demand	Clients are mainly from France, Belgium and Switzerland. They are between 30 and 65 years old They are individual clients or they travel in couple, with their family (20 % of travels includes children), with friends or with an organised group 20 % of travel packages are tailor-made holidays
Motivation for the decision to implement CSR or to specialise in responsible tourism	The motivation for following a responsible path is the result of the founders' desire to translate into business their passion for walking travels carried out with respect for people and places Since its opening in 1986, Atalante has decided to pay attention to the different meanings and implications of travelling: discover of new places, meeting of different people and culture, cultural exchange, support to local development
Policies and actions for the environmental preservation	Among the most important measures, Atalante: – Has signed an agreement with CO_2 Solidaire to compensate the CO_2 emissions produced by employees when they travel by airplane for business purposes – Finances environmental projects in destinations (for example, the installation of solar panels in Cambodia) – Uses recycled paper – Uses eco-friendly products in the office – Uses new energy saving light bulbs – Proposes organic farming and fair trade products to clients – Buys organic farming and fair trade products for its employees
Policies and actions in favour of the local community in destinations	Among the most important measures, Atalante: – Provides training to local providers (for example to guides); – Finances projects for education and wellbeing of local communities in the destination

(continued)

Table 6.2 (continued)

Logo	
Policies and actions towards employees and clients	Among the most important measures in favour of employees, Atalante: – Ensures a fair salary to employees, also reimbursing transport expenses for going to work – Encourages employees to use bicycle or public transports for going to work, if possible – Avoids gender discrimination, by ensuring a fair balance between men and women in the staff – Proposes free wellness care (massages of 15 min a week during working hours) – Promotes training and team building initiatives Among the most important measures in favour of clients, Atalante: – Is extremely concerned with product quality (Atalante is ISO 9001 certified) – Gives transparent and comprehensive information about the destination and the holiday price – Gives client the Traveller Ethic Chart suggesting behaviour and actions to support local community – Encourages clients to voluntary compensate CO_2 emissions – Organises every year trecking events open to clients, partners and employees – For those clients interested in giving their contribution to local community in developing countries, Atalante proposes several practical activities that directly engage tourists
Policies and actions for supporting the economic growth in destinations	Atalante: – Gives priority to local suppliers – Fairly pays local providers and partners with a compensation that is higher than local compensation – Provides training to local providers (for example to guides) – Finances projects aimed at supporting economic growth in destination (for example projects for the development of microcredit)
Reporting system	Since 2008 Atalante is ATR certified. The certification must be updated every 2 years
Projects and investments for the future	Atalante is planning new projects in favour of local communities

Source: authors' reworking on the basis of Atalante' websites (http://www.atalante.fr/).

Table 6.3 SNP Natuurreizen's sheet

Logo	SNP NATUUR REIZEN VAN ANWB
No. of employee and average annual turnover	Employees: 70 Turnover: 28,000,000 €
Main destinations offered by the tour operator	120 destinations across the world: Europe, Asia, North and South America, Antarctic, Africa, Oceania
Average no. of annual clients	Clients: 28,000
Some information about the demand	Clients are all Dutch and Flemish and are on average 57 years old. Group tours account for 33 % of the total demand and free individual travellers for 66 % FIT. Clients travel mostly in couples (especially the individual travellers) or in small group of friends/family or alone (in particular group tours). They book their holidays between 6 months and 1 day before departure
Motivation for the decision to implement CSR or to specialise in responsible tourism	Since SNP's founders like to travel in a responsible way themselves, they have based SNP's philosophy and travel supply on sustainability values. These values are shared by clients, who are interesting in travelling in a way that has a little impact on environment and local culture
Policies and actions for the environmental preservation	SNP tries to lower its impact through many ways as possible. In particular they choose small-scale accommodations (preferably certified), certified nature guides and suppliers, etc
Policies and actions in favour of the local community in destinations	Although the local community is not necessarily involved in the holiday package organisation, SNP works with local agents that follow basic guidelines on social policies
Policies and actions towards employees and clients	SNP provides its employees and tour guides with training courses and also about sustainable tourism principles It gives transparent information to its clients and it asks their opinion and comments after the holiday
Policies and actions for supporting the economic growth in destinations	SNP supports the local community in destinations by working with local small-scale operators and accommodations only
Reporting system	Since 2012, SNP is Travelife certified after going through a very extensive audit procedure. SNP underlines that the audit process is particularly challenging in terms of time and human resources that are needed to filled in the necessary documents The certification must be updated every 2 years
Projects and investments for the future	SNP plans to improve its responsibility

Source: authors' reworking on the basis of information provided by SNP Natuurreizen

and understanding play a relevant role that maybe even more important than the expectation of economic and marketing benefits that CSR can bring to the company.

As regards the environmental and social policies, some initiatives and practices are common to all three tour operators: for example, the compensation of CO_2 emissions, the use of eco-friendly products and recycled paper, the use of public transports, the cooperation with local suppliers in the destination, the guarantee of fair conditions to employees and providers, the attention to product quality and the communication of transparent information to clients.

It is finally interesting to note that the commitment towards responsible tourism and CSR implies a continuous improvement. Tour operators taken in consideration are planning other initiatives and projects to be applied next years.

Chapter 7
Assessing the Effectiveness of Reporting Systems: Why and How

Abstract This chapter proposes and describes a methodology to assess the effectiveness of the reporting systems for responsible tourism and CSR available in Europe, the scope being to give tourism businesses a set of elements useful to choose the program that best responds to their characteristics and requirements. Through the application of a mathematical and multi-criteria decision model, the Analytic Hierarchy Process (AHP), the main characteristics, strengths and weaknesses of every program have been evaluated and compared, according to different criteria. Better than a qualitative analysis, AHP analyses the performance of every alternative with respect to each parameter and finally calculates a value that expresses the overall quality and effectiveness of each system in assessing business responsibility. In particular, the main criteria that have been selected are as follows: the degree of attention paid by every program to all triple bottom line dimensions; the opportunity for small and medium-sized enterprises to apply the system; the integration between the "certification approach" and the "responsible tourism approach"; the transparency given by the type of auditing. The method selects the best reporting system to apply, according to the importance given to each criterion taken into account.

7.1 Why Applying a Mathematical Tool: Main Issues

In Chap. 5, the disparities among reporting systems for responsible tourism and CSR, addressing in particular tour operators, have been analysed and discussed.

The qualitative study has highlighted and compared the main attributes of every program, without however giving any specific detail about the overall quality of the systems themselves, i.e. whether they are effective and appropriate for measuring the business responsibility. This limitation depends in particular on two related elements: firstly, the existence of different criteria, according to which the validity

M. Manente et al., *Responsible Tourism and CSR*, CSR, Sustainability, Ethics & Governance, DOI 10.1007/978-3-319-06308-9_7,

of the programmes should be evaluated; secondly, the evidence that all systems have their strengths and weaknesses, so they can be recommended with regard to some aspects and not for others.

Given these premises, a mathematical and multi-criteria decision model, the Analytic Hierarchy Process (AHP), has been applied in order to study and compare the characteristics of every program according to a set of criteria. The implementation of this model can favour a better understanding of the advantages/disadvantages of each reporting system and then provide tour operators, especially SMEs, with some elements useful to choose the program that best responds to their characteristics and requirements.

The final findings can give some clear guidelines to those—academicians, public opinion, tourism companies—who are looking for more detailed and comprehensive information about the assessment systems. Tour operators, particularly small businesses, can benefit from this support in deciding which reporting program is the most appropriate for their needs.

7.2 The AHP Model: Theoretical and Conceptual Framework

The assessment systems taken in consideration are characterised by many different aspects, both quantitative and qualitative, and it is the integration of these aspects that confers effectiveness and validity to them. Consequently, it is necessary to evaluate the reporting programs in relation to several variables.

The choice should be based on a mathematic model able to assess the reporting systems according to multi-decision criteria—to which different levels of importance can be attributed— and to identify for which criteria each program is more effective than others. The model should consider several information at the same time, compare the programs and combine together the results in order to produce an overall score and then a ranking of the most effective alternatives.

An approach that meets these needs is the Multi-Criteria Decision Analysis (MCDA), a set of a several operation research models that deal with decision problems under the presence of a number of decision criteria (Pohekar and Ramachandran 2004).

MCDA is well known and widely acknowledged in several areas, such as manufacturing systems, investment evaluation, agriculture and environmental management, energy planning, etc. It supports decision-makers by providing solutions when an optimal or at least satisfactory decision is needed under the presence of problems involving conflicting and multiple objective. The decision-making process can be very difficult and compromised by several variables: the presence of quantifiable and non-quantifiable multiple criteria; conflicting objectives; the presence of different groups of decision-makers that are involved in the process and whose preferences highly influence the solution; then the need for a mutual compromise (Pohekar and Ramachandran 2004). Given all these aspects, MCDA can help decision-makers to make a selection among several alternatives or to assign alternatives to different categories or more to establish a ranking or describe the main characteristics of alternatives.

Several methods of MCDA can be identified and, according to a recognised classification proposed in the literature, they can be divided into two main categories: multi-objective decision making (MODM) and multi-attribute decision making (MADM) (Climaco 1997). In multiple-objective decision making, alternatives are not predetermined, but instead a set of objective functions is optimised with respect to a set of constraints in order to find the most satisfactory and efficient solution. On the contrary, in a multi-attribute decision making, a finite number of alternatives are to be evaluated against a set of attributes and the best one is usually selected by making comparisons between alternatives with respect to each attribute (Pohekar and Ramachandran 2004).

Since reporting systems represent a fixed number of options and the main objective of the analysis is to rank them and identify the best one, a multi-attributed model is required.

This approach can be further divided into compensatory and non-compensatory methods, depending on whether they allow trade-offs among attributes. Unlike non-compensatory methods, the use of compensatory techniques allow that poor performance in a number of criteria can be compensated by high performance in other parameters, without reflecting in the aggregated final value (Jeffreys 2004).

Considering that all reporting systems show strengths and weaknesses according to the criteria taken into account, not only the model should be based on a multi-attribute method but also on the compensatory approach.

Among all the currently available methodology, a model that is widely acknowledged and widespread is the Analytic Hierarchy Process (AHP), developed by Thomas Saaty in the 1970s. The model evaluates the overall performance of different alternatives (in this case, the assessment systems), by giving a final score to each individual alternative. The score is obtained by adding together the sum of values achieved by each alternative with respect to every criterion (Saaty 1990a, b, 1991[1]).

The model is recognised as a valid support to the decision-making process and has had considerable success in a variety of fields due to the effectiveness of its approach. Due to its ability to structure and organise the process of analysis that each individual must go through at the moment of making a choice among different alternatives, AHP allows the decision-maker to successfully manage the complexity of the decisional process and to identify the best and most suitable choice.

The reasons behind the decision to adopt AHP not only depends on the opportunity to determine an overall judgment and a ranking of the assessment programs. It provides a guide for tour operators who wish to compare the different assessment systems and decide which one would be the best to use according to their

[1] From now on, unless otherwise indicated, sources are: Saaty 1990a, b, 1991.

characteristics and requirements. In addition, AHP allows the decision-maker to translate qualitative aspects into quantitative assessment terms, relating to the importance given to the different alternatives and criteria. This is an extremely useful feature of the model, since the reporting schemes must be assessed according to both quantitative and qualitative parameters.

The AHP model is based on a 6-step procedure, summarised as follows:

- Identifying the goals to be achieved;
- Defining the criteria and any eventual sub-decision criteria considered to be important for the choice;
- Identifying possible alternative decisions;
- Determining the weights to be assigned to the alternatives in relation to the established criteria;
- Determining the weight of the decision criteria in relation to the goals;
- Building the ranking from the possible alternatives.

The first three phases described above lead towards the development of the dominance hierarchy, i.e. a reticular structure composed of several levels, where the first one (the one at the top) represents the general objective or goal of the assessment, while the second one contains the criteria by which the alternatives are evaluated. Each parameter can in turn be subdivided into more specific sub-criteria and so on, until arriving at the base of the structure, where all different alternatives among which to make a choice are presented. These alternatives are directly linked to more specific parameters (the definitive ones).

As regard the other steps, AHP also implies the calculation of weights (also called priorities) to be assigned to alternatives and criteria. These are numerical values derived from evaluations carried out by experts, which represent, on one hand, the degree of desirability of alternatives with respect to each of the criteria and, on the other, the level of importance of criteria with respect to the objective.

The application of all these steps to the assessment of reporting systems for responsible tourism and CSR is explained in-depth in the following sections.

7.3 Assessing Reporting Programs with the AHP Model: The Process

7.3.1 The Decision Criteria and the Alternatives

As mentioned before, the application of the AHP model has been driven by the objective to assess the effectiveness of the reporting systems available in Europe and then to give tour operators a set of elements useful to choose the program that best responds to their characteristics and requirements. Consequently, the choice of the decision criteria represents a crucial phase in the implementation of the entire model, given that opting for certain parameters rather than others will create significant differences in the final ranking. According to the qualitative analysis carried out in the previous chapters and to the identification of the main characteristics and issues of the reporting systems, four main criteria have been defined (Table 7.1.). These are thought to be the most appropriate variables for assessing reporting programs and determine their effectiveness in measuring the businesses' commitment to responsible tourism and CSR.

The criteria have been divided into sub-criteria in order to better specify some particularly relevant aspects. Table 7.1 shows all the criteria and sub-criteria adopted.

- *Degree of coverage given to all elements of the triple bottom line.* All assessment systems claim to evaluate and certify tour operators' responsibility, but responsibility implies a social, environmental and economic commitment. Therefore, programs that wish to acquire a reputation for seriousness and reliability should be able to assess all the three aspects with appropriate indicators. Consequently, the first criterion judges the thoroughness of each of the methodologies in each of the three areas. The decision was then taken to divide this criterion into three sub-criteria (environmental, social and economic dimension), which in turn were divided into further sub-criteria. Each of the latter represents a category of indicators, whose presence (or absence) contributes to determine the degree of coverage given to the three elements by the system. For example, for the environmental dimension, six sub-criteria were considered, which check the adoption of an environmental policy by the tour operator; the measures implemented to ensure the respect of the environment along the whole value chain; the actions put in place to reduce energy consumption, CO_2 emissions, etc.
- *Degree of implementation of reporting systems on the part of SMEs.* The European tourism industry is mainly composed of small and medium-sized enterprises. The fact that they cannot draw on the same resources as large corporations means that smaller businesses may encounter organisational and economic difficulties with regard to investing in certification programmes. Therefore, the second criterion evaluates systems in relation to their applicability for SMEs. It has been divided into two sub-criteria: the cost of joining and implementing the program and the ease with which the entire process of certification can be carried out by operators (number of indicators to be elaborated and data collection).
- *Level of integration between the certification approach and the responsible tourism approach.* A key element to distinguish the various assessment programmes measuring businesses' responsibility is to identify the approach these programmes follow. As previously explained, the integration of both approaches is crucial in order to gather together simultaneously all aspects of responsibility (strategy, internal processes, activities, travel experiences) and to disseminate pertinent information to all stakeholders, from tourists to the host community, from suppliers to employees. For this reason, the third criterion evaluate systems on the basis of their ability to integrate the two approaches.

Table 7.1 Criteria and sub-criteria for the application of AHP

Criteria 1 Degree of coverage given to all elements of the triple bottom line	
1.1	Environmental dimension: presence of one or more indicators as listed below
1.1.1	Indicators to verify whether the company has an environmental policy in place
1.1.2	Indicators to check the measures adopted by the company with regard to its supply chain and respect for the environment
1.1.3	Indicators inherent to energy consumption, water, paper and the recycling of water, paper and other materials carried out by the company
1.1.4	Indicators with regard to CO_2 emissions produced by the company and the measures adopted by the company to reduce or compensate for the emissions by financing projects to safeguard the environment
1.1.5	Indicators regarding information given out by the company to employees, suppliers, the local community and tourists concerning environmentally friendly behaviour and any environmental problems in holiday destinations
1.1.6	Indicators in relation to the measures adopted by the company to prevent/reduce negative impacts on the environment as a result of tourist activities in holiday destinations (e.g. using local public transport to get around, reducing waste and energy and water consumption, monitoring tourism impact on the environment)
1.2	Social dimension: presence of one or more indicators as listed below
1.2.1	Indicators in relation to the measures adopted by the company to ensure healthy and safe-working conditions for employees as well as offering opportunities for training courses
1.2.2	Indicators with regard to personnel (e.g. the percentage of women in managerial positions, turnover rates, the percentage of job-related accidents and incidents of illness, rate of staff satisfaction)
1.2.3	Indicators in relation to the measures adopted by the company to control whether suppliers have strategies in place to guarantee the safety and well-being of their employees
1.2.4	Indicators aimed at establishing whether the company maintains fair and lasting business relations with their suppliers
1.2.5	Indicators aimed at verifying whether the company acts in the interests of the local community by evaluating existing social problems in the destination and encouraging the local community to be pro-actively involved in the organisation of holidays in the destination
1.2.6	Indicators in relation to the measures adopted by the company to inform tourists about the social and cultural aspects of the destination and whether any meetings or occasions for cultural exchange between tourists and local people are planned and integrated into the holiday package
1.2.7	Indicators regarding measures adopted to ensure clients' safety and satisfaction (questionnaire for the customer satisfaction)
1.3	Economic dimension: presence of one or more indicators as listed below
1.3.1	Quantitative indicators with regard to business performance (e.g. total revenue, percentage of income derived from responsible tourism (if offered), sales of package holidays, main source of income, overheads/expenditure, number of employees, etc.)
1.3.2	Indicators that demonstrate the economic contribution made by the company to the holiday destination in terms of funding development projects and programmes, hiring local people, buying local products and the relationships established with local suppliers (e.g. the percentage of local people employed by the company, the percentage of invoicing to local suppliers, breakdown of the prices of the single components of the holiday package including those inherent to the local community)

(continued)

Table 7.1 (continued)

Criteria 2 Degree of implementation of responsible business practices on the part of SMEs	
2.1	Cost of adhering to and implementing the report
2.2	Simplicity
2.2.1	Number of indicators used by the programme
2.2.2	Ease of data collection and construction of indicators
Criteria 3 Level of integration between the certification approach and the responsible tourism approach	
Criteria 4 Type of auditing (self-assessment, second-party audit and third-party audit)	

- *Type of auditing*. The fourth and final criterion checks the type of auditing adopted by the program: self-evaluation, second-party audit or third-party audit. The choice of the auditing contributes to determine the reliability, objectivity and transparency of a reporting program.

Figure 7.1 shows the dominance hierarchy. The first level at the top represents the general objective of the assessment, while the following three levels contain the criteria and sub-criteria according to which the programs are evaluated. The base of the structure lists the different alternatives, i.e. the reporting systems analysed in Chap. 5. The Responsible Tourism Standards of AITR, the Enterprise Indicator for Responsible Tourism of QUIDAMTUR and the Tour Operators' Sector Supplement of TOI-GRI have been excluded from the analysis, since their principles and statements do not have found a concrete application in the tourism industry.

7.3.2 *The Priorities of Alternatives*

The fourth and fifth phases of the procedure consist in determining the weights (or priorities) of alternatives and criteria.

AHP requires to identify the weights of the alternatives (and of the criteria) by using the technique of paired comparison, which must be carried out by an expert *super partes*. The expert is asked to: compare in pairs all alternatives with respect to each single sub-criterion (and all criteria with respect to the goal); question whether one option is preferable to another (or decide if there is no difference between them) and express a personal judgement using a scale of values developed by Saaty (from 1 to 9). The assessments made by the expert translates into appropriate calculations in order to derive the weights of the alternatives and criteria (see the appendix).

All comparisons of alternatives with respect to criteria and sub-criteria: 1.1.1–1.1.6 (environmental dimension); 1.2.1–1.2.7 (social dimension); 1.3.1–1.3.2 (economic dimension); 2.1, 2.2.1–2.2.2 (implementation by SMEs) and criteria 3 (integration of approaches) and 4 (type of auditing) have been carried out. The assessments have been based on the qualitative analysis described in Chap. 5.

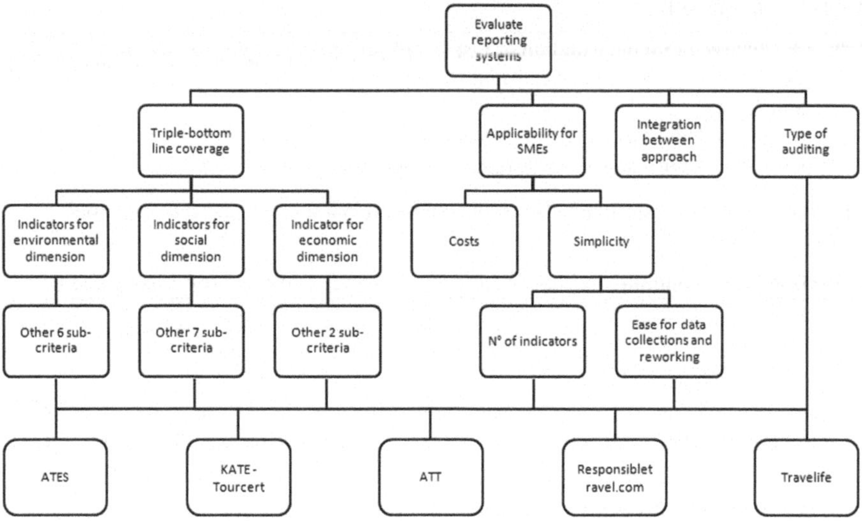

Fig. 7.1 The dominance hierarchy

Before relying completely on the calculated weights, AHP requires to verify if the assessment made in pair comparisons is consistent. It could be happen that the expert encounters difficulty in maintaining coherence in judgment for all the paired comparisons, particularly when there is a high number of alternatives. For example, if alternative A is strongly better than B and B is strongly better than C, it is inconsistent that A and C are equal. Recognising that the assessment may not always be logic, Saaty has fixed a tolerable degree of inconsistency. Referring to the appendix for further explanation about the procedure of the consistency verification, it is however important to note that all comparisons are consistent and therefore that all weights are acceptable.

Tables 7.2, 7.3, 7.4, 7.5, 7.6, and 7.7 show the weights that have been assigned to each alternative/program according to the four criteria and the related sub-criteria.

Considering criterion 2 (implementation by SMEs) and particularly sub-criterion 2.2.1 (number of indicators used), the partial ranking demonstrates that the Responsibletravel.com system obtains the greatest weight (0.48) (Table 7.5). This means that, for the parameter under consideration, it represents the best possible solution for tour operators, in that it requires the implementation of the lowest number of indicators in comparison to others. On the contrary, the Travelife Sustainability System shows the lowest weight, given the high number of indicators businesses have to build in order to accomplish with its procedure.

As for the type of auditing (criterion 4), the programs of KATE-Tourcert, ATT and Travelife are the best solution (weight equal to 0.31), since both these systems require a third-party audit (Table 7.7).

The same reasoning can be followed for other criteria and sub-criteria in order to understand what are the best systems in terms of coverage of the environmental,

Table 7.2 Weights of reporting systems with respect to sub-criteria 1.1.1–1.1.6 (environmental dimension)

	1.1.1	1.1.2	1.1.3	1.1.4	1.1.5	1.1.6
A	0.03	0.03	0.05	0.24	0.07	0.13
B	0.24	0.34	0.43	0.24	0.07	0.13
C	0.24	0.19	0.05	0.24	0.24	0.13
D	0.24	0.10	0.05	0.03	0.07	0.04
E	0.24	0.34	0.43	0.24	0.56	0.56

A. ATES
B. KATE-TourCert
C. ATT
D. Responsibletravel.com
E. Travelife

Table 7.3 Weights of reporting systems with respect to sub-criteria 1.2.1–1.2.7 (social dimension)

	1.2.1	1.2.2	1.2.3	1.2.4	1.2.5	1.2.6	1.2.7
A	0.10	0.08	0.24	0.33	0.58	0.49	0.24
B	0.24	0.69	0.24	0.05	0.09	0.04	0.24
C	0.04	0.08	0.24	0.51	0.09	0.31	0.24
D	0.04	0.08	0.03	0.05	0.04	0.08	0.03
E	0.57	0.08	0.24	0.05	0.21	0.08	0.24

Table 7.4 Weights of reporting systems with respect to sub-criteria 1.3.1–1.3.2 (economic dimension)

	1.3.1	1.3.2
A	0.08	0.29
B	0.69	0.29
C	0.08	0.08
D	0.08	0.05
E	0.08	0.29

Table 7.5 Weights of reporting systems with respect to sub-criteria 2.1 and 2.2.1–2.2.2 (implementation by SMEs)

	2.1	2.2.1	2.2.2
A	0.40	0.17	0.21
B	0.07	0.07	0.06
C	0.07	0.25	0.21
D	0.40	0.48	0.47
E	0.07	0.03	0.06

Table 7.6 Weights of reporting systems with respect to criterion 3 (integration of approaches)

	3
A	0.19
B	0.04
C	0.48
D	0.09
E	0.19

social and economic dimensions (Tables 7.2, 7.3, and 7.4) or of adhesion and implementation costs (Table 7.5) or of integration between certification approach and responsible tourism approach (Table 7.6).

Table 7.7 Weights of reporting systems with respect to criterion 4 (type of auditing)

	4
A	0.03
B	0.31
C	0.31
D	0.03
E	0.31

7.3.3 The Weights of Criteria: Different Hypothesis

Before building the final ranking of reporting programs, a further step aims at determining the priorities among the decision criteria and their respective sub-criteria in relation to the objective of the analysis, again through the technique of paired comparison. Although all parameters play an important role in evaluating the alternatives, some of them may be considered to be more influential than others.

Determining the different weights to be assigned to criteria represents an extremely crucial phase in that their value can influence the final classification. For example, a programme can get a low score in one particular criterion. The more significant that parameter is considered to be, the greater the impacts on the programme's final position in the ranking.

As regards the sub-criteria identified for each criterion, the same importance was assigned to them (i.e. the same weights). The basic assumption is that their contribution to the analysis is similar. For example, subscription costs (2.1) and ease of implementation (2.2), identified within criterion 2, were given a weight of 0.50 each.

Considering instead the four main criteria (coverage of triple bottom line; implementation by SMEs; integration of approaches and type of auditing), five hypothesis are proposed in order to demonstrate how the weights of each alternative/program in the final ranking can change depending on the allocation of the priorities to criteria. Each assumption shows a different distribution of weights and then a different way to perceive the relative importance of criteria (Table 7.8). The different distribution of weights among criteria for each hypothesis is summarised in Table 7.8, which shows that in the last four assumptions not only the order of importance of criteria is different but also the weight value. In detail:

- The first hypothesis assigns the same weight to all four criteria (0.25), assuming that they equally contribute to assess the reliability of a reporting system;
- The other assumptions test different allocations of priorities, stating that the criteria have a different impact on programs' ranking. For example, the second hypothesis gives greatest importance to the first criterion (coverage of the triple bottom line), the third hypothesis to the second criterion (applicability for SMEs), while the fourth one to the third criterion (integration of approaches and the last one to the fourth parameter (type of auditing).

Table 7.8 Weights of four main criteria—different hypothesis

	Hypothesis 1	Hypothesis 2	Hypothesis 3	Hypothesis 4	Hypothesis 5
Criterion 1	0.25	0.54	0.30	0.23	0.25
Criterion 2	0.25	0.25	0.53	0.23	0.25
Criterion 3	0.25	0.05	0.11	0.44	0.15
Criterion 4	0.25	0.16	0.06	0.10	0.35

Table 7.9 Weights of reporting programs with respect to four main criteria

	1	2	3	4
A	0.190	0.293	0.193	0.034
B	0.320	0.067	0.041	0.310
C	0.159	0.149	0.483	0.310
D	0.068	0.436	0.091	0.034
E	0.263	0.056	0.192	0.310

Note: for better highlighting differences among programs, weights are rounded to the third decimal place

Table 7.10 Final weights of reporting systems—different hypothesis

	Hypothesis 1	Hypothesis 2	Hypothesis 3	Hypothesis 4	Hypothesis 5
A	0.177	0.191	0.235	0.199	0.162
B	0.185	0.241	0.155	0.138	0.212
C	0.275	0.197	0.198	0.314	0.258
D	0.157	0.156	0.264	0.160	0.152
E	0.205	0.215	0.148	0.189	0.217

Note: for better underlining differences among programs, weights are rounded to the third decimal place

7.3.4 *The Final Ranking*

In order to determine the overall importance of each of the alternatives (in our case, the reporting systems) and to draw up a final ranking, according to the different hypotheses made, AHP requires to multiply each of the weights attributed to the individual alternatives (Tables 7.2, 7.3, 7.4, 7.5, 7.6, and 7.7) by the weights of each sub-criteria and by those of each criteria (Table 7.8) and then to add all the results together.

While the weights of the reporting programs with respect to the third and fourth criteria are already known, since they have no sub-criteria, those referring to the first and second parameters have to be calculated as sum of the products among the partial weights of alternatives and the weights of the sub-criteria.

The weights of the reporting systems according to the four main criteria are summarised in Table 7.9. The final ranking of programs, according to the five hypotheses made, is described in Table 7.10, which shows how priorities are different on the basis of the different allocation of weights among criteria.

7.4 How to Choose the Best Reporting Program for Responsible Tourism and CSR: Main Findings

The previous Table 7.10 summarises the final ranking of alternatives and highlights how a different allocation of weights among decision criteria influences the judgement on the effectiveness and quality of reporting programs.

Following the first hypothesis (equal weight to all criteria), the most effective system is that developed by ATT, which achieves the highest weight (0.275). The same applies when the fourth and fifth hypotheses are taken into account: when the highest importance is given to criterion 3 (implementation by SMEs) or to criterion 4 (type of auditing) respectively, ATT confirms its supremacy (weights equal to 0.314 and 0.258). Instead following the second hypothesis, which assigns the highest value to the first criterion (coverage of the triple bottom line), the most reliable scheme is the CSR Reporting Standards of KATE-TourCert, which obtains a weight of 0.241. This means that this scheme covers all three dimensions of responsibility (environmental, social and economic) better than other systems do (Table 7.9). Finally, if the second assumption is considered (implementation by SMEs) the ResponsibleTravel.com's program is the best alternative (weight equal to 0.264), which implies that it can be adapted to the specific requirements of small and medium-sized businesses more easily than others.

In general, the ATT system seems to be the most reliable reporting program, according to three out of five hypotheses made. This is not a surprise since, as shown in Table 7.9, it reaches the highest scores in two out of four criteria and satisfactory weights in the other two parameters. More specifically, it adequately covers the three dimensions of responsibility with a sufficient set of indicators; it can also be applied by SMEs; it appropriately integrates the responsible tourism approach with the certification one and it is transparent, thanks to a third-party assessment. This does not mean that ATT is the "ideal" scheme, but that it is able to combine all aspects on which the effectiveness of a program has been based, better than the other alternatives.

AHP has the advantage of highlighting the main attributes, strengths and weaknesses of every alternative, summarising the overall quality of each system. Not only the final ranking provides tour operators with an overall judgment on the reporting programs available, but also the partial priorities indicate how systems perform according to the decision criteria taken into account. From this point of view, the application of this mathematical model also stresses the key role that the selection of criteria and the decision about their importance play in the process and definition of systems' quality and effectiveness.

Apart from these advantages, there are some issues that need to be taken into consideration, in particular whether new reporting programs appear on the market and need to be added to the assessment process. Firstly, paired comparisons could become difficult for the expert, in particular when the number of alternatives (i.e. the programs to be evaluated) is high. The same applies when the number of criteria and sub-criteria used increases. In this situation, the procedure is likely to become both more demanding and time consuming and, since it may impact negatively on the evaluation consistency, it requires to pay greater attention in the evaluation of alternatives. However, in this specific case study comparisons have proved to be not only simple to be carried out but also consistent.

Another issue is that whenever a new alternative/program is added to the assessing procedure, the performance of "old" programs should be re-assessed, even when these programmes have not been modified. This because it is not possible to determine the weights of the new entrant leaving the other options untouched. The reason behind this problem is obviously linked to the procedure on which the AHP model is based: the priorities are determined by comparing each option with all the others, i.e. making relative comparisons, not absolute assessments. Consequently, the paired comparisons have to be recalculated, as well as the weights of all alternatives. However, this aspect can also be seen as an advantage of the AHP model: not only each program is evaluated according to its specific features, but also to the "business" context in which it operates, i.e. the attributes of the other schemes available on the market. This context can change over time, according to the evolution of the programs already proposed and to the development of new programs by other institutions and bodies.

Generally speaking, the application of AHP provides a useful and updated tool to tourism companies engaged in assessing the responsibility of their tourism activities and internal practices and their CSR. It can support them in evaluating and certifying their business responsibility and, at the same time, impact on business reputation and consumers' trust.

Appendix: Some Details About the AHP Procedure

This appendix proposes a further explanation of the AHP model, in particular of the calculation of weights, starting from the pair comparison technique, and of the consistency verification.

Determining the weights of the alternatives signifies assigning to them a numerical value that expresses their suitability and importance compared to the other options for each of the criteria. AHP requires to assign priorities starting from the technique of *pair comparison*, in other words, by comparing two by two all the alternatives, with respect to each single sub-criterion, and questioning whether one is preferable to another and to what extent.

The operation must be entrusted to experts familiar with the subject, who will be asked to express their personal preferences according to a scale developed by Saaty:

- 1 If alternative i and alternative j are equal;
- 3 If alternative i is moderately preferable to alternative j;
- 5 If alternative i is strongly preferable to alternative j;
- 7 If alternative i is more strongly preferable to alternative j;
- 9 If alternative i is extremely preferable to alternative j;
- 2,4,6,8 As intermediate values (or in the case of compromise).

The result of the comparison is the coefficient of dominance a_{ij}, an estimate of the dominance of the first element i) compared to the second (j). In particular, making a paired comparison of n elements results in n^2 coefficients, of which only n $(n-1)/2$ must be directly determined by the expert who is carrying out the evaluation, in that:

$$a_{ii} = 1, \tag{7.1}$$

and

$$a_{ji} = 1/a_{ij} \tag{7.2}$$

for each of the values of i and j.

The coefficients of dominance determine therefore a positive reciprocal square matrix with n rows and n columns (where n is equal to the number of alternatives) called *matrix of paired comparisons*.

Below a detailed explanation of all steps of the procedure is proposed, taking as example the comparison of all five reporting systems with respect to sub-criterion 2.2.1 (number of indicators). From the paired comparisons, a 5×5 matrix is obtained (Table 7.11), where the values along the main diagonal correspond to 1, given that each alternative is equal to itself. Considering, for example, the comparison between C and E, it is possible to see that C is extremely preferred to E, since it has a lower number of indicators, and that the coefficient of dominance of E compared to C is none other than the reciprocal of that of C compared to E. The other comparisons can be interpreted in the same way.

At this point, the AHP model requires to:

- Standardise the matrix of paired comparisons, calculating the sum of each of the columns and dividing each element of the matrix by the value of the sum of the vector column to which that particular element belongs (Table 7.12);
- Calculate the average of each row of the standardised matrix, obtaining in this way the vector of weights w (Table 7.13).

Before relying completely on the calculated weights, it is necessary to verify the *consistency* of the original matrix of paired comparison. AHP requires to:

- Multiply the matrix of the paired comparisons (not standardised) by the vector of weights w (Table 7.14);
- Divide this vector by the vector of the weights w, to obtain the vector λ_{max} (Table 7.15);
- Calculate the average of the values that make up vector λ_{max} in order to obtain an average λ_{max}.

When the matrix is perfectly consistent, $\lambda_{max} = n$, i.e. to the number of alternatives that must be compared.

By calculating the average of the values that make up vector λ_{max}, we obtain a λ_{max} average equal to 5.27. If the matrix were perfectly consistent, $\lambda_{max} = n = 5$. In

Table 7.11 Matrix of paired comparisons—sub-criteria 2.2.1

	A	B	C	D	E
A	1	3	0.5	0.25	8
B	0.33	1	0.2	0.17	4
C	2	5	1	0.33	9
D	4	6	3	1	9
E	0.13	0.25	0.11	0.11	1

Table 7.12 Standardised matrix of paired comparison—sub-criteria 2.2.1

	A	B	C	D	E
A	0.13	0.20	0.10	0.13	0.26
B	0.04	0.07	0.04	0.09	0.13
C	0.27	0.33	0.21	0.18	0.29
D	0.54	0.39	0.62	0.54	0.29
E	0.02	0.02	0.02	0.06	0.03
Sum	1	1	1	1	1

Table 7.13 Vector of weights—sub-criteria 2.2.1

	w
A	0.17
B	0.07
C	0.25
D	0.48
E	0.03

Table 7.14 Matrix of paired comparison multiplied by the vector of weights—sub-criteria 2.2.1

0.87
0.38
1.38
2.61
0.15

Table 7.15 Vector $\lambda_{\max}$

5.27
5.10
5.42
5.49
5.06

our case, this does not happen, but if we calculate the two indices specifically developed by Saaty, the Consistency Index (CI) and the Consistency Ratio (CR), we can consider it to be acceptable. Indeed if $CR \leq 0.10$, the degree of inconsistency in the paired comparisons matrix is acceptable and it is therefore possible to assume that the priorities obtained are significant. If on the other hand $CR > 0.10$, it may be that serious inconsistencies exist in the paired comparisons, and as a consequence, the analysis may not bring relevant results.

In this case, with $CR = 0.06$, the matrix may be considered to be consistent and the weights calculated to be significant.[2]

$$CI = (\lambda_{max} - n)/(n - 1) = 0.07 \tag{7.3}$$

$$CR = CI/1.12 = 0.06 < 0.10 \tag{7.4}$$

References

Climaco J (ed) (1997) Multicriteria analysis. Springer, New York

Jeffreys I (2004) The use of compensatory and non-compensatory multi-criteria analysis for small-scale forestry, small-scale forest economics. Manage Pol 3(1):99–117

Pohekar SD, Ramachandran M (2004) Application of multi-criteria decision making to sustainable energy planning - A review. Renew Sustain Energy Rev 8:365–381

Saaty TL (1990a) The analytic hierarchy process: planning, priority setting, resources allocation. RWS Publications, Pittsburgh

Saaty TL (1990b) Decision making for leaders: the analytic hierarchy process for decisions in a complex world. University of Pittsburgh, Pittsburgh

Saaty TL (1991) Prediction, projection and forecasting: applications of the analytic hierarchy process in economic finance, politics, games and sports. Kluwer, Boston

[2] 1.12 in (7.4) is the value of the Random Index for $n = 5$; the random index represents the value that the Consistency Index would assume in the hypothesis that all the elements of the matrices of paired comparisons were chosen in a casual manner, all the elements of the main diagonal would be equal to 1 and $a_{ji} = 1/a_{ij}$.

Chapter 8
Discussion and Future Research

Abstract This chapter firstly discusses the main implications of the analysis of reporting systems for responsible tourism and CSR. In particular, it provides tour operators with detailed information on every available scheme and its performance/benefits in order to stimulate their adoption and to support more informed decisions about the system that best responds to their specificities and requirements. The study also enables students and researchers to build or enhance their knowledge about the characteristics and development of the main reporting initiatives available in Europe and to assess the potential of the AHP model for this kind of study. In the second part, future research directions are debated: the opportunity to analyse other reporting systems, developed in other world regions or addressed to other tourism companies (not only to tour operators); the adoption of other criteria for comparing and evaluating systems; the integration of reporting programs with demand-side assessment tools, such as online user-generated contents.

8.1 Implications of the Reporting Systems' Analysis

The presence in the European context of several and heterogeneous initiatives for assessing the responsibility of tourism businesses and the opacity of most programs make it difficult for operators, consumers and society in general to have access to comprehensive information about these schemes and to be aware of their effects and impacts.

In addition, few studies have been concerned with analysing reporting systems that are specifically focused on business responsibility and addressed in particular to tour operators. The literature has proposed several contributions on the diffusion of responsible tourism and CSR practices in the tourism industry and on the issues

M. Manente et al., *Responsible Tourism and CSR*, CSR, Sustainability, Ethics & Governance, DOI 10.1007/978-3-319-06308-9_8,

met by tourism companies in following a responsible path, but it has paid less attention to describe the main reporting programs and identify their characteristics.

Therefore, the qualitative analysis and the application of the AHP model developed in this book contribute to enhance the body of knowledge about reporting systems. In particular, this innovative approach represents a valid tool for comparing and assessing the reporting programs according to several different criteria and for identifying the most reliable solutions.

The study can represent a support for both the academic community and the tourism industry. On the one hand, students and researchers can build or improve their knowledge about: the characteristics and development of the main reporting initiatives available in Europe; the most committed organisations and relevant approaches; the main variables affecting the overall quality and effectiveness of these systems. In addition, the analysis helps them to assess the potential of the AHP model, for this kind of study or for other related studies, and can support further research in this field, for example the comparison of reporting systems developed outside Europe or addressed to other tourism companies.

On the other hand, tourism operators can understand the benefits that the implementation of responsible policies and activities and the adoption of suitable reporting systems bring to their business, in terms of efficiency improvement, cost reduction, visibility, brand image and consumer trust. In detail, the analysis provides them with detailed information on every scheme available in Europe and, through the final ranking obtained with the AHP model, on its strengths and weaknesses in order to stimulate their adoption and to support more informed decisions about the system that best responds to their specificities and requirements.

The results show that currently there is not an "ideal" solution, which satisfies all decision criteria. Every system has its advantages and disadvantages and then can be a valid option according to some aspects or for some operators and not for others. Therefore, the selection of the most suitable and effective program depends on the criteria that are considered to be a priority: the coverage of the triple bottom line, the implementation by SMEs, the integration of approaches and the type of auditing. For example, a micro enterprise can take its decision according to the ranking of programs obtained following the third hypothesis, which attributes the highest weight to the second criterion (i.e. the applicability of the system by SMEs: see Tables 7.8 and 7.10),

It should be noted that not all systems included in the analysis are open to all tour operators. Those developed by ATES and ATT have been developed for the members of these associations, while the Responsibletravel.com program addresses partners and providers of the online travel agency. However, even if not members, tour operators can take inspiration from these schemes, learn about the indicators used, the minimum standards and the assessment procedure and then adapt these systems to their specific situation.

Another issue that is interesting to highlight concerns the identification of the "expert" who, according to the AHP model, is in charge for the definition of the criteria and sub-criteria and, particularly, for the weights to be assigned to the reporting programs with respect to criteria, and to criteria according to the objective. This role can be taken by academicians, like in this analysis, or by a *super partes* organisation, for example those associations committed to promote responsible tourism and CSR and to support, through several initiatives, tourism companies in implementing responsible practices.

8.2 Future Research Directions

The methodology applied to describe and review the reporting systems available in Europe proved to be effective in assessing the performance of each scheme according to a set of predefined criteria and to provide useful guidelines to researchers an operators. However, further research is required to test it with larger set of programs and/or different criteria.

From the first point of view, attention should be paid to other initiatives and experiences developed in the international context. For example, in South Africa and South America tourism organisations are particularly dynamic and motivated in encouraging tourism industry to operate according to responsible principles (Greening the WSSD Initiative 2003; Bien 2008). Future studies may verify whether and to what extent reporting programs have been developed; what are their main characteristics and the differences with European systems; whether the issues discussed in this chapter are also significant in other countries. Starting from Goodwin, who argues that reporting systems' variety reflects the cultural and ecological diversity of cultures and countries, it would be interesting to check how diverse values, traditions and practices influence the approaches and the standards on which reporting systems are based.

A correlated issue concerns the comparison with reporting systems developed for tourism companies (such as accommodation, travel agencies, transports) along the tourism value chain or for destinations. Although some basic indicators are applicable to all enterprises (see, for example, the GRI scheme, which is based on the Sustainability Reporting Guidelines common to all business sectors), there is the need to customise programs according to the type of business. This because companies can differ in terms of internal processes, relationships with suppliers, employees and consumers, impacts on local community and environment. This explains why two organisations considered in the analysis—Travelife and KATE-Tourcert—have developed two different versions of the same reporting system: one scheme for tour operators and another one for accommodation. These versions share the same approach, but differ in terms of type of indicators, minimum standards, auditing procedure. Therefore, further research could study and compare reporting systems addressed to different businesses of the tourism supply chain.

Another area of further improvement, related to the previous research directions, may check the feasibility of integrating systems developed by different organisations and applied in different countries or in different companies along the tourism value chain. This opportunity could indeed encourage an assessment and certification of all activities and services included in the tourism package. An example is the mutual recognition agreement recently signed by TourCert and Travelife with Fair Trade Tourism in order to support their members and to stimulate cooperation among reporting systems. As recognized by these organisations, similar initiatives can provide reliable reporting and auditing services to companies and improve the recognition and the awareness of reporting schemes within the tourism industry.

From the point of view of the decision criteria and sub-criteria used in the model, their selection was based on a deep study of programs' characteristics and issues. Therefore, they can be considered significant and effective in judging system's quality. However, future studies can check the opportunity to integrate other criteria and analyse the differences in partial and final rankings produced by their inclusion. For example, a new criterion can be represented by the relationship between responsible practices and consumer trust. As discussed in Chap. 3, the adoption of responsible policies and activities and the certification of this commitment through a reporting system positively impact on consumer confidence. However, the company engaging in responsible tourism or in CSR needs clear evidence that its efforts have contributed to enhance consumer trust. Therefore, a reporting scheme that not only assesses business responsibility but also measures the impacts in terms of clients' satisfaction and opinions, increase in the number of responsible packages sold, etc. would be really valuable. Some reporting systems considered in the study include these aspects, for example encouraging the company to submit a customer satisfaction questionnaire to clients (e.g. KATE-TourCert). Not only the results of this questionnaire contribute to assess the business responsibility in terms of respect of consumer rights; they also give the business useful indications about how travel packages are perceived by clients, the importance of responsible products for their purchase and consumption and their intentions to buy other products in the future.

A final area for future research concerns the investigation of other tools for monitoring the business responsibility, different from reporting programs. In the era of Web 2.0, the growing participation of consumers to social media and their interest to tell their own travel experience can indirectly impact on reporting systems. Due to the development of new technologies and platforms, social media enable people to generate their own contents (user-generated contents) to interact with companies and express their positive or negative judgement about products and activities. While in the past travellers used social media mainly during their decision-making process, for selecting and learning more about the destination, nowadays they are more active in generating contents also during and after the travel, posting comments, reviews, photos and videos (see, for example, Gretzel et al. 2007).

The reviews and judgments posted by online users on tourism destinations, products and companies represent a tool that support other travellers in choosing their holiday resort, airline, hotel, etc. but can also integrate reporting and certification systems (Manente et al 2011). Generally, consumers are more likely to trust other consumers who have already experienced the product (word of mouth effect) rather than certifications and reporting systems, even when they have been developed by an independent third-party audit (Honey 2005; Kennedy 2012; Sparks et al 2013). A similar phenomenon can take place in the field of responsible tourism and CSR: tourists can become active in telling their responsible travel, communicating their satisfaction and giving feedbacks about the company based on their concrete experience. For example, they can express their comments about the respect shown by the company for the environment and the local culture, the use of local employees and suppliers, the ability of the guide in stimulating a real awareness and culture exchange, etc. (Manente et al 2011). This phenomenon then leads to a system based on user-generated content, where the auditing is carried out by the market.

The importance of this topic is proved by some initiatives considered in this study that have already followed this road. For example, ResponsibleTravel.com largely bases its scheme on the verified comments posted on the website by tourists who have already experienced the packages sold in the online travel agency. Another similar initiative, even if limited to the environmental dimension only, is the TripAdvisor Green Leaders program. Launched in April 2013 by the world's largest travel site, the program supports travellers in their green choices by highlighting eco-friendly hotels through a four level ranking—bronze, silver, gold and platinum—. In the future, the system will also invite online users to comment on a hotel's green credentials following a stay at a participating property, and their feedback will be displayed on the site.

References

Bien A (2008) A simple user's guide to certification for sustainable tourism and ecotourism. CESD—Center on ecotourism and sustainable development handbook, n. 1

Greening the WSSD Initiative (2003) Responsible tourism handbook. A guide to good practices for tourism operators. http://www.capetown.gov.za/en/tourism/Pages/ResponsibleTourism.aspx. Accessed 23 Oct 2013

Gretzel U, Yoo KH, Purifoy M (2007) Online travel review study. Laboratory for intelligent systems in tourism. http://www.tripadvisor.com/pdfs/OnlineTravelReviewReport.pdf. Accessed 15 Oct 2013

Honey M (Ed) (2005) Marketing strategy for sustainable tourism certification. CESD - Center on Ecotourism and Sustainable Development, The International Ecotourism Society

Kennedy R (2012) Responsible tourism and social media marketing. In: Goodwin H, Font X (eds) Progress in responsible tourism, vol 2. Goodfellow Publishers Limited, Oxford, pp 4–33

Manente M, Minghetti V, Mingotto E (2011) Turismo responsabile e CSR: guida e confronto tra programmi di valutazione per uno sviluppo sostenibile. Franco Angeli, Milano

Sparks BA, Perkins HE, Buckley R (2013) Online travel reviews as persuasive communication: the effects of content type, source, and certification logos on consumer behaviour. Tourism Manage 39:1–9

Chapter 9
Conclusion

Abstract In the last chapter, some conclusions are drawn by summarising the most significant findings of the study. The book represents an interesting contribution to the ongoing debate about the role of responsible tourism, CSR and related reporting programs to encourage tourism companies and travellers to do their part for sustainable development in tourism.

Given the issues that limit the diffusion of responsible policies and practices in the tourist industry, numerous initiatives have been developed to raise awareness of tourism businesses about the importance to join the principles of responsible tourism and to implement CSR strategies. Reporting systems represent a tool that can encourage and support tourism operators in following a responsible path.

Not only this book provides a comprehensive review of some important aspects referring to responsible tourism and CSR and specifically of reporting systems. It also aims at being a stimulus for positive reflection and discussion, offering some contributions to the ongoing debate.

In particular, the manuscript has focused its attention on the significant role that responsible tourism and CSR can play in promoting sustainable development in tourism and in contributing to the competitiveness of tourism companies. The compliance with responsible tourism principles and the integration of CSR in the core business can promote a positive balance between the company's goals, based on profit, and the needs of other stakeholders, who are directly or indirectly linked to the company and influence its performance: employees, providers, consumers, host community and society in general. In addition, the book has discussed how responsible policies and activities positively contribute to build and retain consumer trust. From this point of view, also small businesses can strengthen their position by encouraging consumers to trust in their products.

For those readers arguing that the adoption of responsible behaviour does not bring positive impacts on companies, it has been highlighted that in tourism, even more than in other sectors, responsible practices are of vital importance. Not only

M. Manente et al., *Responsible Tourism and CSR*, CSR, Sustainability, Ethics & Governance, DOI 10.1007/978-3-319-06308-9_9,

embracing sustainability principles impacts positively on environment and society; it also contributes to increase the quality and to add value to tourism products, ensuring in this way the competitiveness of companies and of the overall tourism industry in a destination.

Secondly, the book has analysed the reporting systems that are currently available in Europe to assess the business responsibility of tourism companies, in particular of tour operators. The decision to concentrate on reporting systems is based on the crucial importance that these tools play both for businesses and consumers. On the one hand, they support the company in reorganising its business and in improving its responsibility from a social, economic and environmental point of view. On the other hand, they support consumers in taking responsible decisions about their holidays, by increasing their awareness about responsible companies and products and giving transparent assurance about companies' commitment.

The application of both a qualitative analysis and a quantitative study, through the application of the AHP model, has highlighted the great variety of programs developed in Europe. Although all programs aim at assessing tour operators' activities and their impacts, they differ in terms of the approach chosen, the indicators selected, the minimum standards identified, the auditing procedure applied, the attention paid to the triple bottom line dimensions and their applicability to SMEs.

In the presence of a very opaque and heterogeneous context, this book offers some clear guidelines for both the academic community and the tourism industry, merging theoretical and practical implications. Not only it enriches the body of knowledge about reporting systems, but it also proposes tour operators a methodology to compare them and to identify the most effective one according to their characteristics, priorities and requirements.

All findings aim at being a stimulus for future research that can support further analysis of these systems and their broad diffusion and, indirectly, the spread of responsible tourism and CSR practices.

The manufacturer's authorised representative in the EU is Springer Nature Customer Service Centre GmbH, Europaplatz 3, 69115 Heidelberg, Germany. If you have any concerns regarding our products, please contact ProductSafety@springernature.com

Printed and bound by CPI Group (UK) Ltd, Croydon, CR0 4YY
05/10/2026
02228154-0004